Hydrogen Sulfide

The project that is the subject of this report was approved by the Governing Board of the National Research Council, whose members are drawn from the Councils of the National Academy of Sciences, the National Academy of Engineering, and the Institute of Medicine. The members of the Committee responsible for the report were chosen for their special competences and with regard for appropriate balance.

This report has been reviewed by a group other than the authors according to procedures approved by a Report Review Committee consisting of members of the National Academy of Sciences, the National Academy of Engineering, and the Institute of Medicine.

The work on which this publication is based was performed pursuant to Contract No. 68-02-1226 with the Environmental Protection Agency.

Hydrogen Sulfide

Subcommittee on Hydrogen Sulfide

Committee on Medical and Biologic Effects of Environmental Pollutants

Division of Medical Sciences
Assembly of Life Sciences
National Research Council

University Park Press
Baltimore

UNIVERSITY PARK PRESS
International Publishers in Science and Medicine
233 East Redwood Street
Baltimore, Maryland 21202

Typeset by American Graphic Arts Corporation.
Manufactured in the United States of America by The Maple Press Company.

Library of Congress Cataloging in Publication Data

National Research Council. Committee on Medical and Biologic Effects of Environmental Pollutants. Subcommittee on Hydrogen Sulfide.
Hydrogen sulfide.

(Committee on medical and biologic effects of environmental pollutants)
Bibliography: p.
Includes index.
1. Hydrogen sulphide—Toxicology. 2. Hydrogen sulphide—Physiological effect. 3. Hydrogen sulphide—Environmental aspects. I. Title.
[DNLM: 1. Environmental pollution. 2. Hydrogen sulfide—Toxicity. QV662 N277h]
RA1247.H8N37 1978 615.9′1 78-14561
ISBN 0-8391-0127-9

Contents

Committee on Medical and Biologic Effects of Environmental Pollutants

Reuel A. Stallones University of Texas, Houston, Texas, *Chairman*

Martin Alexander Cornell University, Ithaca, New York

Andrew A. Benson University of California, La Jolla, California

Ronald F. Coburn University of Pennsylvania School of Medicine, Philadelphia, Pennsylvania

Clement A. Finch University of Washington School of Medicine, Seattle, Washington

Eville Gorham University of Minnesota, Minneapolis, Minnesota

Robert I. Henkin Georgetown University Medical Center, Washington, D.C.

Ian T. T. Higgins University of Michigan, Ann Arbor, Michigan

Joe W. Hightower Rice University, Houston, Texas

Henry Kamin Duke University Medical Center, Durham, North Carolina

Orville A. Levander Agricultural Research Center, Beltsville, Maryland

Roger P. Smith Dartmouth Medical School, Hanover, New Hampshire

T. D. Boaz, Jr. Division of Medical Sciences, National Research Council, Washington, D.C., *Executive Director*

Subcommittee on Hydrogen Sulfide

Roger P. Smith Dartmouth Medical School, Hanover, New Hampshire, *Chairman*

Robert C. Cooper University of California, Berkeley, California

Trygg Engen Brown University, Providence, Rhode Island

E. R. Hendrickson Environmental Science and Engineering, Inc., Gainesville, Florida

Morris Katz York University, Downsview, Ontario, Canada

Thomas H. Milby Environmental Health Associates, Inc., Berkeley, California

J. Brian Mudd University of California, Riverside, California

August T. Rossano University of Washington, Seattle, Washington

John Redmond, Jr. Division of Medical Sciences, National Research Council, Washington, D.C., *Staff Officer*

Preface

In 1924, Mitchell and Davenport[220] prepared a review and a supplemental reference list that summarized a large body of early obscure literature on hydrogen sulfide. This fascinating and succinct analysis, which covers 150 years of scientific observations on the subject, remains remarkably clear and penetrating even by modern standards. It has been reprinted in this monograph as Appendix II so that the reader will have access to reports published prior to the end of the nineteenth century. Accordingly, in the text of this monograph citations to this very early literature have been held to a minimum.

This report marks the bicentennial of the first systematic study of the preparation and properties of hydrogen sulfide, which was published in 1777 by the Swedish chemist, Carl Wilhelm Scheele. Scheele reported that an odorous gas resulted from the action of mineral acids on certain inorganic sulfides. The same gas could be prepared by heating sulfur in the presence of hydrogen. He made the first observations on the solubility of the gas in water and on its oxidation to sulfur by a variety of agents. Scheele called the gas *Schwefelluft* (sulfur air) or referred to it more prosaically as *stinkende* (stinking or fetid). Like many chemists, Scheele had little appreciation for the violently poisonous nature of the materials with which he worked. He had already discovered hydrogen cyanide. (The significance of this coincidence will be apparent later.) Many historians have remarked that he was fortunate to have escaped with his life.[247]

Coincident with Scheele's discovery of hydrogen sulfide, a series of accidental exposures to sewer gas in Paris resulted in many deaths. Thus, in a certain sense, the toxicological history of hydrogen sulfide began in the sewers of Paris. Forty years were to elapse, however, before hydrogen sulfide was retrospectively implicated as the causative agent. An even earlier description of the effects of hydrogen sulfide on the eyes of cesspit and privy cleaners can be found in the remarkable treatise on occupational health by Bernardino Ramazzini first published in 1713.[255]

Any mention of the sewers of Paris is certain to bring to mind the best-known novel of the most important French Romantic writer. Victor Hugo's *Les Miserables* first appeared in 1862.[142] In one of its memorable scenes, Jean Valjean bore the unconscious Marius toward the sewer outlet where the relentless Inspector Javert was waiting.

Since Hugo was known as a careful researcher, his fascinating account of the history of the sewer system of Paris carries with it the weight of authority. His morbid preoccupation with the subject was commonly shared by the public. According to Hugo, this could be traced far back into the history of human waste disposal:

> The sewers and drains played a great part in the Middle Ages, under the Lower Empire and in the old East. Plague sprang from them and despots died of it. The multitudes regarded almost with a religious awe these beds of corruption, these monstrous cradles of death. The vermin-ditch at Benares is not more fearful than the Lion's den at Babylon. (p. 199)[142]

It is tempting to speculate that part of the fear was due to a general appreciation of the toxicity of hydrogen sulfide, which might have been empirically recognized since very early times. Hugo may have been thinking of the 1777 accidents when he compared death in quicksand with death in the sewer:

> Slow asphyxia by uncleanliness, a sarcophagus where asphyxia opens its claws in the filth and clutches you by the throat; fetidness mingled with the death-rattle, mud

> instead of the sand, sulphuretted hydrogen in lieu of the hurricane, ordure instead of the ocean! (p. 250)[142]

To Hugo the sewer was the "Intestine of the Leviathan." That analogy may be better than he imagined since some sources insist that hydrogen sulfide is produced in the human bowel as well. The implications of endogenous hydrogen sulfide production and many other things remain unknown about this chemical.

Acknowledgments

This document was prepared by the Subcommittee on Hydrogen Sulfide under the chairmanship of Dr. Roger P. Smith. Although the initial drafts of the various sections were prepared by individual subcommittee members, the entire document was extensively reviewed by the entire subcommittee and represents a group effort.

Chapter 1 contains a review of the occurrences, properties, and uses of hydrogen sulfide. This material was prepared by Dr. E. R. Hendrickson. In Chapter 2, the biogeochemical aspects of the sulfur cycle are discussed by Dr. Robert C. Cooper.

The Preface and Chapters 3 and 4 were written by Dr. Smith. In Chapter 3, he describes absorption, distribution, metabolism, and excretion of sulfides in animals and humans. Chapter 4 contains a summary of the experimentation that has been done on the effects of hydrogen sulfide in animals. The author is grateful for the 15 years of research support provided by the Public Health Service for some of the studies reported in these chapters.

Dr. Thomas H. Milby wrote Chapter 5, in which the effects of hydrogen sulfide on humans are examined. A discussion of the effects on vegetation and aquatic animals follows in Chapter 6, a contribution from Dr. J. Brian Mudd.

Chapter 7 concerns the establishment of air quality standards or criteria for hydrogen sulfide. Dr. August T. Rossano provided this material.

Dr. Trygg Engen discusses both the psychological and aesthetic aspects of odor in Chapter 8. The odor of hydrogen sulfide is one of the most well recognized characteristics of the gas.

Chapters 9 and 10, which contain the summary and conclusions, and the subcommittee's recommendations, respectively, were assembled by Dr. Smith from material supplied by the subcommittee.

Appendix I, which is a review of sampling and analysis techniques, was written by Dr. Morris Katz.

The preparation of the document was assisted by the comments from anonymous reviewers designated by the Assembly of Life Sciences and from members of the Committee on Medical and Biologic Effects of Environmental Pollutants. Dr. Robert J. M. Horton of the Environmental Protection Agency gave invaluable assistance by providing the subcommittee with various documents and translations. Information assistance was obtained from the National Research Council Advisory Center on Toxicology, the National Academy of Sciences Library, the Library of Congress, the Department of Commerce Library, and the Air Pollution Technical Information Center.

The staff officer for the Subcommittee on Hydrogen Sulfide was Mr. John Redmond, Jr. The references were verified and prepared for publication by Mrs. Louise Mulligan, Ms. Joan Stokes, and Ms. Ute Hayman. The document was edited by Mrs. Frances M. Peter.

Hydrogen Sulfide

1

Hydrogen Sulfide—Its Properties, Occurrences, and Uses

PROPERTIES

There are a number of hydrogen sulfides, including polysulfides and hydrosulfides; however, hydrogen sulfide (H_2S) is the most common. Hydrogen sulfide is a colorless gas having the characteristic odor of rotten eggs. The gas is flammable, burning in air with a pale blue flame. The ignition temperature is 260 C. Mixtures in air between 4.3% and 46% by volume hydrogen sulfide are explosive.[335]

Hydrogen sulfide is a liquid at minus 61.8 C and a solid at minus 82.9 C.[205] The specific gravity of the gas is 1.189 when the specific gravity of the air is taken at 1.00. One liter of hydrogen sulfide at 0 C and 760 mm weighs 1.5392 g.[335] The vapor pressure at various temperatures is shown in Table 1-1.

Hydrogen sulfide is soluble in amine solutions, in alkali carbonates, bicarbonates, and hydrosulfides, in hydrocarbon solvents, in ether, in alcohol, in glycerol, in water, and in several other solvents. Water solutions are not stable, inasmuch as the absorbed oxygen causes the formation of elemental sulfur and the solutions become turbid quite rapidly.

A number of agents are capable of oxidizing hydrogen sulfide. The rate of reaction and the compounds that are formed depend mainly on the oxidizing agent, its concentration, and the conditions of the reaction. Some of the oxidation reactions are summarized in Table 1-2.

DISTRIBUTION

Hydrogen sulfide occurs naturally in coal, natural gas, oil, volcanic gases, and sulfur springs and lakes. It is also a product of the anaerobic decomposition of sulfur-containing organic matter. In these natural occurrences, other sulfur compounds are nearly always present with the hydrogen sulfide.

Natural Sources

In the United States, natural gas deposits are among the richest natural sources of hydrogen sulfide. Particularly large and important deposits are

Table 1-1. Vapor pressures of hydrogen sulfide at various temperatures[a]

Temperature (C)	Vapor pressure (atm)
0	10.8
10	14.1
20	18.5
30	23.6
40	29.7
50	36.5
60	44.5
70	53.1
80	64.0
90	72.6
100	88.7

[a] From Macaluso, 1969.[205]

located in central and north-central Wyoming, in western Texas, in southeastern New Mexico, and in Arkansas. Hydrogen sulfide concentrations as high as 42% are present in the gas from central Wyoming, where the reserves of that gas are estimated to be about 59 billion kg.[102]

The sulfur content of petroleum deposits in the United States varies from about 0.04% in Pennsylvania crude oil to about 5% in Mississippi crude oil.[205] The sulfur in petroleum consists entirely of divalent sulfur compounds of carbon and hydrogen.

Coal deposits in the United States contain sulfur mainly in pyrites or sulfate. The sulfur content ranges from a trace to more than 8%. The hydrogen sulfide that is occasionally encountered in coal mining operations probably results from the action of steam on the pyrites at high temperatures.

The recovery of hydrogen sulfide in particular and sulfur in general from these fuels has grown dramatically since the mid-1950's. This increase has resulted from efforts to clean up the source material, not from operations to recover the gas for its value. Recently, air quality standards have contributed to the increased sulfur recovery from fuels. The economics of these recovery operations are more favorable with certain rich sources, such as sour natural gas and some refinery gas streams. Worldwide recovery of hydrogen sulfide was estimated in 1965[44] to exceed 5 million metric tons. (See Table 1-3.) Most hydrogen sulfide recovered from fuel is converted to high quality sulfur that must compete with sulfur from natural sources in the same market. These recovery processes are described in this chapter under "Recovery of Hydrogen Sulfide," p. 6.

The production of hydrogen sulfide from volcanic gases is attributed to the action of steam on inorganic sulfides at high temperatures. Similar action is probably responsible for the hydrogen sulfide content of the steam from geothermal "wells." In sulfur springs and lakes, which occur in a variety of locations, hydrogen sulfide is probably produced by both chemical reactions and bacteriologic decomposition of mineral sulfates. Under anaerobic conditions, bacteriologic decomposition of protein and other sulfur-containing organic matter is responsible for the familiar odor of hydrogen sulfide.

Table 1-2. Oxidation reactions involving hydrogen sulfide[a]

Oxidizing agent	Conditions	Chief products
Oxygen (air)	Flame, air in excess	Sulfur dioxide
	Flame, hydrogen sulfide in excess	Sulfur
	Aqueous solution of hydrogen sulfide	Sulfur
Sulfur dioxide	Elevated temperature, catalyst	Sulfur
	Aqueous solution	Sulfur, polythionic acids
Sulfuric acid	Concentrated acid	Sulfur, sulfur-dioxide
Hydrogen peroxide	Neutral solution	Sulfur
	Alkaline solution	Sulfurous acid, sulfates
Ozone	Aqueous solution	Sulfur, sulfuric acid
Nitric acid	Concentrated aqueous solution	Sulfuric acid
Nitric oxide	Silica gel catalyst	Sulfur
Nitrogen dioxide	pH 5 to 7	Sulfur, nitric oxide
	pH 8 to 9	Sulfur, ammonia
Chlorine	Gaseous reaction, excess chlorine	Sulfur dichloride
	Gaseous reaction, excess hydrogen sulfide	Sulfur
	Aqueous solution, excess chlorine	Sulfuric acid
Iodine	Aqueous solution	Sulfur
Iron (Fe^{3+})	Aqueous solution	Sulfur

[a] From Macaluso, 1969.[205]

Table 1-3. Worldwide recovery of hydrogen sulfide in 1965[a]

1965 plant recovery capacities per year (million metric tons)		Amount recovered in 1965 (million metric tons)	
By country		By source	
United States	1.95	Natural gas	4.0
Canada	2.45	Oil refineries	1.0
France	1.52	Coal gas and other	0.2
Other	0.68	Total	5.2
Total	6.60		
By source		By form	
Natural gas	4.7	Elemental sulfur	4.9
Oil and coal	2.0	Other, chiefly sulfuric acid	0.3
Total	6.7	Total	5.2

[a] From Macaluso, 1969.[205]

Generally, no attempts are made to control emissions of hydrogen sulfides from these natural sources. However, greater consideration of these emissions may be required when the tapping of geothermal energy sources increases as projected.

Industrial Sources

Hydrogen sulfide is a by-product of or waste material from a number of industrial operations. Wherever sulfur or certain sulfur compounds come into contact with organic materials at high temperatures, hydrogen sulfide could be formed. The gas also may be released in the course of sulfur recovery operations. It frequently mixes with other odorous sulfur compounds.

In the production of carbon disulfide, sulfur is reacted with natural gas at elevated temperatures. Half of the sulfur introduced is consumed in the production of hydrogen sulfide. In the older process for producing carbon disulfide, the sulfur vapors were reacted with charcoal. This produced smaller, but nevertheless appreciable, quantities of hydrogen sulfide.

Depending on the sulfur content of the basic raw material, substantial quantities of hydrogen sulfide can be released during the production of coke or of manufactured gases from coal. The coal is heated, then quenched in water. The production of steam causes the release of hydrogen sulfide from mineral sulfides. In the case of manufactured gas, the hydrogen sulfide is an undesirable impurity that is usually removed by passage through boxes of iron oxide.[205] Control of hydrogen sulfide emissions from the coking operation is very difficult and is almost never practiced. Similar emissions of hydrogen sulfide may result from the manufacture of reactive petroleum coke.

The process for making thiophene requires the reaction of sulfur with butane at elevated temperatures. This reaction also produces hydrogen sulfide.[205]

Several steps in the refining of petroleum products require the removal and recovery of sulfur compounds. These recovery operations are described in this chapter under "Recovery of Hydrogen Sulfide," p. 6.

In the manufacture of viscose rayon, cellulose pulp is treated with sodium hydroxide, then with carbon disulfide, and again with sodium hydroxide to produce the viscose solution. The viscose solution may be used to spin fibers or for coating other materials. The viscose solution after spinning or coating is passed through a series of acid coagulation baths where hydrogen sulfide is released. From 6 to 9 kg of hydrogen sulfide are formed per 100 kg of rayon produced.[205]

Another major industrial source of hydrogen sulfide is the kraft process for producing chemical pulp from wood. Hydrosulfides are used in the woodchip cooking liquor. After the cooking, which is done at elevated temperatures and pressure, the spent cooking liquor is evaporated and burned to recover the cooking chemicals and heat. Thus, hydrogen sulfide, along with other odorous sulfur compounds, is released at every major process step, including the recovery furnace, direct-contact evaporator, digester, multiple-effect evaporator, oxidation towers, brown stock washers, smelt tank, and lime kiln. Hydrogen sulfide is generally the largest gaseous emission from the kraft process. A number of techniques have been developed to reduce the emission of these odorous sulfur compounds. Depending on the process steps involved and the control techniques that are applied, emissions of hydrogen sulfide can range from less than 0.5 kg to more than 20 kg of gas per 1,000 kg of air-dried pulp produced.

INDUSTRIAL USES

Most of the hydrogen sulfide that is recovered from the sources described above is converted to elemental sulfur or sulfuric acid. It may first be converted to elemental sulfur, then later at a different location be used to manufacture sulfuric acid. If there is a substantial market for sulfuric acid in the vicinity of the recovery process, sulfuric acid may be produced directly without going through the intermediate step of sulfur production. Elemental sulfur, however, is a convenient form for shipping and storage. Processes for recovering hydrogen sulfide and converting it to elemental sulfur or other compounds are described in this chapter under "Recovery of Hydrogen Sulfide," p. 6.

Hydrogen sulfide also is used to prepare various inorganic and organic sulfur compounds. Large quantities are used directly in the manufacture of sulfide, sodium hydrosulfide, and organic sulfur compounds such as thiophenes, thiols, thioaldehydes, and thioketones. Hydrogen sulfide has been reacted with various organic reagents in the development of extreme pressure lubricants and cutting oils. Sometimes it is also used to remove arsenic from the sulfuric acid that is produced from pyrites. The treatment removes not only the arsenic but also other heavy metal impurities. Thus, acid produced in this manner is relatively pure.

The leather industry uses substantial amounts of sodium sulfide in the wet operations of preparing hides for tanning. After a preliminary cleaning with water, hides are placed in pits containing solutions of calcium hydroxide with or without sodium sulfide. This treatment loosens the hair at the roots and permits its mechanical removal. On sheepskins, sodium sulfide is applied as a paste, and the wool can be pulled off on the following day. Actual tanning is subsequently accomplished by soaking the hides in solutions of basic chrome sulfate.[76]

Ton quantities of hydrogen sulfide are used in some installations for the production of heavy water, which can serve as a moderator in nuclear power reactors. Operational plants are located in Aiken, South Carolina and in Glace Bay, Nova Scotia. The advantage of heavy water as a nuclear moderator is that it permits reactor operation with natural uranium instead of the more expensive enriched fuel.

The heavy water is produced in a dual temperature process in which water extracts deuterium from hydrogen sulfide in cold (30 C) towers and hydrogen sulfide extracts deuterium from water in hot (130 C) towers. The exchange is ionic, and it occurs in the liquid phase. After a number of stages and distillations, deuterium oxide of 99.8% purity can be produced.[169]

RECOVERY OF HYDROGEN SULFIDE

Conversion of Sulfur Compounds in Petroleum to Hydrogen Sulfide

As indicated previously, all of the sulfur that is found in crude oil consists of divalent sulfur compounds of carbon and hydrogen. These are principally thiols, sulfides, thiophenes, and benzothiophenes. It is convenient to convert these to hydrogen sulfide prior to the production of other useful sulfur compounds. Using by-product hydrogen from the catalytic reforming operations of the refinery, the sulfur compounds are converted to hydrogen sulfide in the presence of a catalyst. The hydrogen sulfide is absorbed from the gas stream and, subsequently, is desorbed as by-product hydrogen sulfide. About 80% to 90% of the sulfur compounds are converted in this process.[205] The commercial processes for this conversion differ only in the nature of the catalysts used and whether the catalyst is on a fixed or a fluidized bed.

The Unifining process utilizes a fixed-bed reactor containing a cobalt molybdate catalyst. After the feedstock is mixed with hydrogen, it is placed in the reactor at about 370 C. After several intermediate steps the desired hydrogen sulfide is removed by stripping. In the Shell hydrodesulfurization (HDS) process, a reactor is packed with a cobalt-molybdenum-aluminum catalyst or a tungsten-nickel sulfide catalyst. The Gulf HDS process resembles the Shell process except that its temperatures are higher and its space velocities are lower. Also, a fractionation step is used to separate the two fractions in place of the more usual stripping. Other similar processes are the H-oil fluidized-bed process and the fixed-bed Isomax process.

Recovery Processes

There are a number of processes for the recovery of the hydrogen sulfide that is produced by the methods described above or that occurs directly in fuels or industrial off-gases.[205] These processes can be broadly classified as absorption-desorption and oxidation to oxides or elemental sulfur.

The absorption-desorption processes utilize either alkaline liquids or organic solutions as absorbents. They all have the same basic flow scheme. The gas containing the hydrogen sulfide is introduced into the bottom of a reaction chamber or absorber, and the absorbent flows counter to the gas flow. The gas leaving the top of the absorber is essentially free of hydrogen sulfide, which has been transferred to the absorbent. The solution from the bottom of the absorber is pumped to the top of a reactivating tower where it flows counter to a flow of steam produced by boiling the solution in the bottom of the tower. The steam rising through the solution strips the hydrogen sulfide. The gas stream is then condensed to recover the hydrogen sulfide. The stripped solution is sent back into the absorber for re-use.

One of the processes most widely used to recover hydrogen sulfide from natural and refinery gases is the Girbotol process. The absorbent used in this process is usually an aqueous solution of monoethanolamine or diethanolamine. A relatively small amount of steam is required to strip the amine solution to a very low concentration of hydrogen sulfide. Since the ethanolamine is readily contaminated with materials such as tars, the process has not been used to clean the hydrogen sulfide concentration from manufactured gas.

The Shell phosphate process is similar to the Girbotol process but uses absorbent solutions containing over 40% of tripotassium phosphate. This process has the advantage that live steam can be used for stripping and hydrogen sulfide can be selectively absorbed in the presence of carbon dioxide.

The Koppers vacuum carbonate process is a modification of an earlier process developed by the same company. The absorbent solution is sodium carbonate. To conserve steam, stripping is carried out under a vacuum. This process can be used to remove hydrogen sulfide from manufactured gas.

The Shell Sulfinol process has come into commercial use within the last 10 years. The absorbent known as "Sulfinol" is composed of sulfolane plus an alkanolamine. The advantage of this absorbent over the more usual ethanolamines is the higher capacity to remove hydrogen sulfide. The absorbed hydrogen sulfide can be stripped by using smaller quantities of steam at lower temperatures.

Other widely used industrial processes do not recover the hydrogen sulfide as such. Some of these processes are used as a final clean-up operation after one of the aforementioned processes has been used. Among these is the so-called "dry-box" process, which is used to clean coke oven gases. Hydrated iron oxide, which is coated on shavings or other support material, serves as the contactor. The hydrogen sulfide reacts to form ferric sulfide, which, with added oxygen, is reoxidized to the original iron oxide and sulfur. Although the dry-box process is highly efficient in removing hydrogen sulfide, the sulfur cannot be recovered. Where very complete removal of small concentrations of hydrogen sulfide is required, such as in some industrial off-gases, a sodium hydroxide scrubber is used. In a two-stage scrubbing installation it is possible to produce sodium hydrosulfide, which may be concentrated and sold.

Oxidation of Hydrogen Sulfide to Sulfur

As mentioned above, hydrogen sulfide is often converted to elemental sulfur to facilitate shipping and storing if there is no market for sulfuric acid near the recovery operation. There are two groups of such conversion processes—those that are used mainly with sources rich in hydrogen sulfide, such as sour natural gas and refinery gas streams, and those used in industrial processes such as viscose rayon manufacturing.[205]

For high volume gas streams that are rich in hydrogen sulfide, the basic technique for sulfur production is the Claus process. Essentially, a portion of the hydrogen sulfide is burned to sulfur dioxide, which is then combined with the remaining hydrogen sulfide in the presence of a catalyst to produce elemental sulfur. More frequently used is the Stanolind-modified Claus process, in which one-third of the acid gas feed is burned with stoichiometric volumes of air to form sulfur dioxide. This is combined with the remaining hydrogen sulfide in the presence of a catalyst to produce elemental sulfur. Depending on whether a one-stage or a two-stage unit is used, 80% to 92% of the theoretical conversion is possible. The sulfur is condensed out of the gas stream as liquid sulfur.

In more dilute gas streams the hydrogen sulfide may be absorbed in a solution; then, with a suspension of catalyst in the solution, it is oxidized to sulfur by the air that is present. The Ferrox process uses a suspension of iron oxide as the catalyst. The nickel process uses nickel sulfate. In the so-called Thylox process a neutral solution of sodium thioarsenate is used instead of sodium carbonate. In this case the thioarsenate serves as both the absorbent and the catalyst. In all three processes, the absorbent with its suspended catalyst contacts the gas stream containing hydrogen sulfide, as previously described. The absorbent is then regenerated in a tall contact tower. The absorbent, along with compressed air, is introduced into the bottom of the tower. As the air bubbles travel up the tower, they oxidize the sulfur. The sulfur particles are carried upward with the bubbles and a froth of sulfur is skimmed from the top of the tower. This sulfur is not as pure as that produced by the Claus process.

2

Biogeochemical Aspects of the Sulfur Cycle

Microorganisms are frequently involved in the translocation and transformation of minerals of all kinds. These biogeochemical activities may be associated either indirectly or directly with the metabolism of the microorganisms. In a number of instances both direct and indirect activities are involved.

Indirect biogeochemical activities include the dissolution of minerals due to the acidic conditions that result from microbial metabolism, the precipitation of minerals produced by reducing conditions, the adsorption of minerals to microbial surfaces, and the formation and destruction of organometallic complexes.

In the transformation by direct metabolic activity, the minerals are either trace elements in the cellular apparatus or serve as specific oxidizable substrates, electron donors, or electron acceptors in the oxidation-reduction activities of microbial metabolism. As trace elements, minerals do not exhibit any obvious mass movement; however, their involvement in the other activities may bring about significant mineral transformations.

There are three types of the general oxidation-reduction reactions involving minerals:

Type I. Reduced minerals (MH_2) are oxidized by autotrophic or mixotrophic microorganisms. The energy derived from the oxidation is utilized in cell synthesis, and an oxidized form of the mineral (M) is produced:

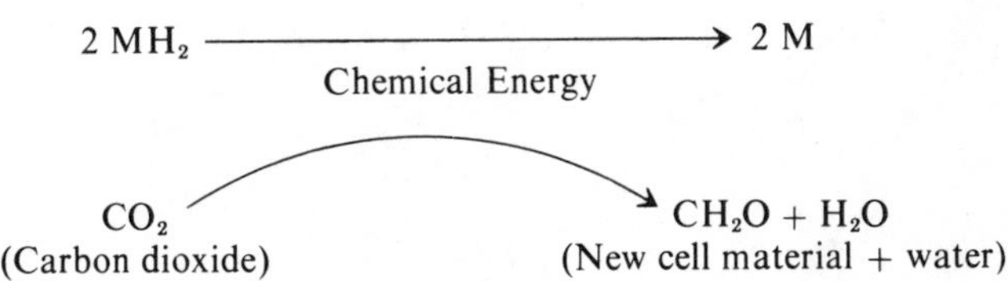

Type II. A reduced mineral acts as an electron donor for bacterial photosynthetic activity. Light energy drives the reaction:

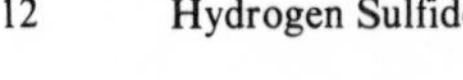

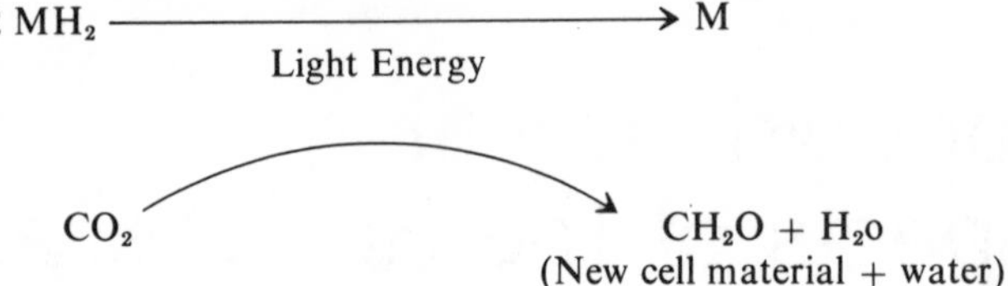

Type III. Oxidized minerals act as electron acceptors for heterotrophic and mixotrophic bacteria:

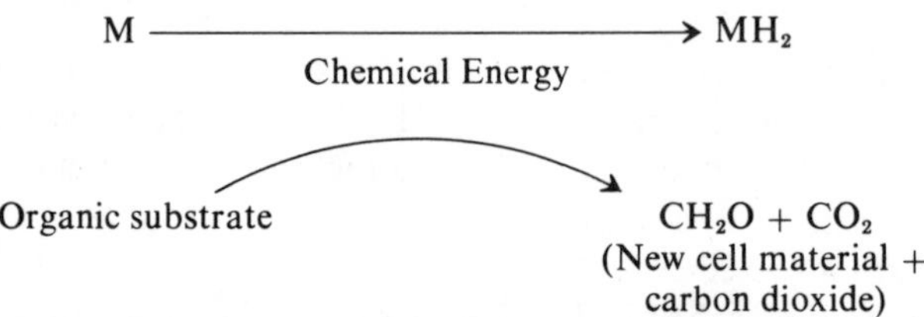

Sulfur is the major mineral element involved in all three of the above oxidation-reduction reactions. Frequently, hydrogen sulfide is the reduced mineral form. In the sulfur cycle, shown diagrammatically in Figure 2-1, these three microbial reactions play a prominent role.

Sulfur, which is cycled throughout the environment, involves diverse types of microorganisms. Hydrogen sulfide is oxidized to elemental sulfur by a variety of bacteria found in soil and water. These bacteria include a number of filamentous forms, such as members of the genera *Beggiatoa*, *Thioploca*, and *Thiothrix*. Of these, the genus *Beggiatoa* is studied the most. These bacteria grow in straight filaments of less than 1

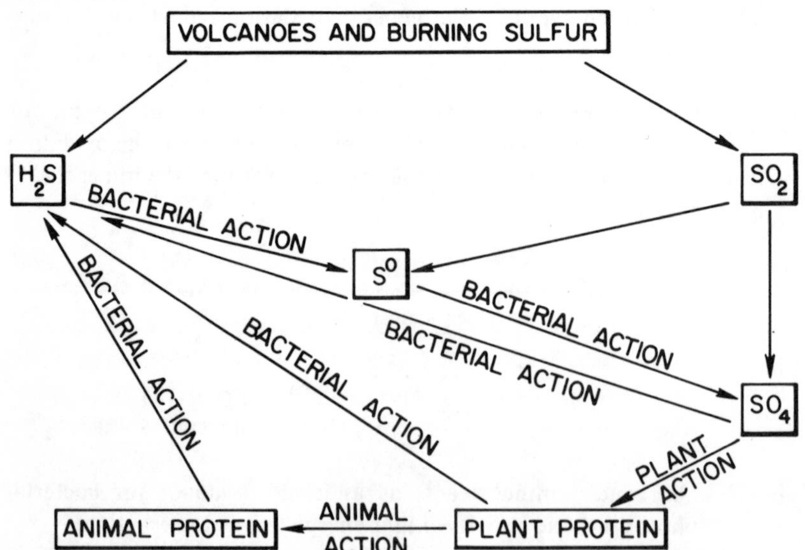

Figure 2-1. The sulfur cycle. From Cooper *et al.*, 1976.[74] Reprinted with permission.

to 55 μm in width, exhibit a creeping motility, and, in the presence of hydrogen sulfide, deposit elemental sulfur inside their cells. The metabolism of all these filamentous organisms from the three genera is respiratory, using molecular oxygen as the terminal electron acceptor. Ecologically, they function in transition zones between aerobic and anaerobic conditions where molecular oxygen and hydrogen sulfide can exist simultaneously.

Beggiatoa not only are important in the biogeochemical formation of sulfur but also appear to play an important role in the protection of plant species that are adversely affected by hydrogen sulfide in the soil. In rice paddies, an accumulation of hydrogen sulfide can inhibit oxygen production and nutrient uptake by the rice plants. The presence of *Beggiatoa* is beneficial in that these microorganisms will oxidize hydrogen sulfide, thus allowing the rice plants to flourish.[157] In the mud in Lake Ixpaco, Guatemala, where hydrogen sulfide is of volcanic origin, *Beggiatoa* have been found in numbers as high as $100{,}000/cm^3$.[200]

A number of other genera of bacteria are associated with environments in which oxygenated water interfaces with overlying water containing hydrogen sulfide. These genera include *Thiobacterium*, *Macromonas*, *Thiovulum*, and *Thiospira*.[50] There are no reports of these organisms having been isolated in pure culture. Consequently, their interaction with hydrogen sulfide has not been firmly established.

Photosynthetic bacteria belonging to the families Chromatiaceae and Chlorobiaceae all oxidize hydrogen sulfide to elemental sulfur and sulfate in the presence of light. Their metabolic pattern is described in the Type II oxidation-reduction reaction (p. 11). It can be more specifically represented by the following equation:

$$CO_2 + 2H_2S \xrightarrow[\text{Photosynthetic bacteria}]{\text{Light}} \underset{\text{(New cells*)}}{CH_2O^*} + 2S^0 + H_2O \qquad (1)$$

Genera in both families are anaerobic and obligately phototrophic. They are found in waters that contain relatively high concentrations of hydrogen sulfide. The purple sulfur bacteria of the family Chromatiaceae appear to be primarily photoautotrophic. The microorganisms in these families include a variety of morphologic types, such as rods, spirals, and spheres, and a variety of forms of cell aggregation, from single cells to cubical masses like those in the genus *Sarcina*. They differ from eukaryotic green plants in the chemical variety of their chlorophyll; in their basic photosynthetic process, which is classified as cyclic photophosphorylation; and in the wavelength of light required for metabolic activity. Colorful blooms of these organisms, particularly the purple sulfur bacteria, occur in highly sulfurous bodies of water such as cyrenaic lakes,[179] where hydrogen sulfide is actively oxidized by these bacteria, thereby producing elemental sulfur. As much as 200 tons of this sulfur is produced each year.[53, 54] Heavy growths of purple sulfur bacteria also occur in anaerobic waste-treatment lagoons. They effectively oxidize the hydrogen sulfide that is produced.[73] Under laboratory conditions these microorganisms could oxidize up to 300 mg/liter of hydrogen sulfide/day.

Reduced sulfur compounds are also oxidized in nature by members of the bacterial genus *Thiobacillus*. There are eight species recognized in *Bergey's Manual of Determinative Bacteriology*.[50] All are gram-negative rods and, with one exception, are obligate aerobes. Five of the species are strict autotrophs, two are facultative autotrophs, and one (*T. perometabolis*) is heterotrophic, but requires simultaneous utilization of sulfur compounds and organic substrates for optimal growth. The majority of these species oxidize hydrogen sulfide, lower oxides of sulfur, and elemental sulfur. One species, *T. ferrooxidans*, is also able to oxidize iron. The end result of their oxidative activity is the production of sulfate, which tends to create acidic conditions. The thiobacilli are noted for their tolerance to low pH conditions. *T. thiooxidans* and *T. ferrooxidans* have growth optima at pH 2.5 and 4, respectively.[340] *T. ferrooxidans* are frequently involved in the production of acid mine wastes, which are initiated by the oxidation of iron pyrite.[204, 313]

Silverman and Ehrlich[279] propose that iron pyrite (FeS_2) reacts with oxygen and water to produce acidic (H_2SO_4) waters:

$$FeS_2 + 3.5O_2 + H_2O \rightarrow FeSO_4 + H_2SO_4 \tag{2}$$

Reaction 2 occurs spontaneously in the presence of oxygen and can also be effected by some thiobacilli. Ferrous sulfate ($FeSO_4$) normally oxidizes to ferric sulfate [$Fe_2(SO_4)_3$] slowly; however, in the presence of *T. ferrooxidans* the rate of oxidation is 0.1 to 1 million times faster.[313] The following equation summarizes the reaction:

$$4FeSO_4 + 2H_2SO_4 + O_2 \rightarrow 2Fe_2(SO_4)_3 + 2H_2O \tag{3}$$

In the presence of ferric sulfate, iron pyrite is inorganically oxidized to ferrous sulfate:

$$FeS_2 + Fe_2(SO_4)_3 \rightarrow 3FeSO_4 + 2S \tag{4}$$

This reaction creates more substrate for the population of *T. ferrooxidans*, with resulting oxidative activity, more dissolution of the iron pyrite, and production of sulfuric acid. The elemental sulfur produced during these activities will be available for further oxidation by microorganisms such as *T. thiooxidans* to form more sulfuric acid. Acid mine drainage, an important environmental problem in some areas of the United States, is the result of these microbiologic activities. An autotrophic, nonsulfur-oxidizing bacterium, *Ferrobacillus ferrooxidans*, has been frequently reported to be active in the production of acid mine water;[185] however, the latest edition of *Bergey's Manual*[50] includes this genus in the description of *Thiobacillus ferrooxidans*.

Brock *et al.*[45] described a new genus of sulfur-oxidizing bacteria, *Sulfolobus*, which resemble microorganisms from the genus *Mycoplasma* and oxidize elemental sulfur to sulfate in natural acidic habitats such as those found in geothermal areas of Yellowstone National Park. These aerobic organisms are thermophilic, with an optimum growth temperature of 70 to 75 C. Their optimum pH is 2 to 3. The authors suggest that this group of bacteria may be important in the production of sulfuric acid from sulfur in high-temperature hydrothermal systems. These organisms are not known to oxidize hydrogen sulfide.

The action of the microorganisms discussed thus far relates to the oxidation of hydrogen sulfide to elemental sulfur and, ultimately, to sulfate. Once formed, sulfate is extremely stable to further chemical activity in nature. It is reduced essentially through biologic processes and, to a large extent, through the direct activities of bacteria.

In the sulfur cycle (Figure 2-1) sulfate is reduced to hydrogen sulfide indirectly through uptake by plants and incorporation into plant protein. These plant proteins are incorporated into animal protein by herbivorous animals, and subsequently progress through the food web continuum. The decay of plant and animal material through bacterial action results in the production of hydrogen sulfide and completes the cycle. Direct reduction of sulfate to hydrogen sulfide is brought about by specialized, strictly anaerobic, sulfate-reducing bacteria.

A large array of organic compounds can act as the source of hydrogen sulfide for an equally large array of heterotrophic microorganisms. The organic compounds involved are those that contain sulfur-bearing amino acids, such as proteins, peptides, and glutathione.[10] Many bacteria, fungi, and actinomycetes release hydrogen sulfide to the environment during the decay of these compounds. An example of a common bacterium that produces hydrogen sulfide in the presence of protein is the heterotroph *Proteus vulgaris*. Hydrogen sulfide is sometimes noticeably generated from organic sources such as those in sewage treatment plants and in solid waste disposal sites. For example, sulfide concentrations as high as 24.8 mg per liter in a sewage stabilization pond have been reported.[10] In this instance, the air 15 meters from the pond contained between 6.7 and 8.8 ppm.

Other volatile sulfur compounds are also released to the atmosphere as a result of microbial activity in the degradation of organic matter. These compounds include: sulfite, carbonyl sulfide, methanethiol, dimethyl sulfide, dimethyl disulfide, and ethanethiol, propane, and butanethiols.[10, 159] Dimethyl sulfide, which is found in sea water, may be a significant component of the sulfur cycle, particularly the sulfur flux between sea and land.[202] Direct methylation of sulfate by wood-rotting fungi to form methanethiol has also been reported.[36]

The microbiologic reduction of sulfate to hydrogen sulfide is accomplished by members of two genera of anaerobic bacteria: *Desulfovibrio* (five species) and *Desulfotomaculum* (three species). These bacteria are all gram-negative, strictly anaerobic, heterotrophic, and have a respiratory metabolism in which sulfates, sulfites, or other reducible sulfur compounds serve as the final electron acceptors, with the resultant production of hydrogen sulfide (see metabolic Type III). The organic substrates for these organisms are usually short chain acids, such as lactic and pyruvic acid. In nature, these substrates are provided through the fermentative activities of anaerobic bacteria on more complex organic material. Thus, when oxygen is depleted, organic material is present, and sulfate is available, one could expect the production of copious amounts of hydrogen sulfide. For example, a heavy rainstorm (45.7 cm in 24 hr) on the island of Oahu, Hawaii, inundated an extinct volcanic crater in which dredge spoils from Pearl Harbor were stored. Prior to the storm, the crater had been seeded in an attempt to control windblown dust. A succulent plant known locally as "Akulikuli grass" abounded. The resulting biogenic hydrogen sulfide reached levels as high as 20 ppm in the surrounding air.[122]

The activities of these sulfate-reducing bacteria are important, not only in the production of hydrogen sulfide *per se* but also because of the interaction of sulfides with other materials in the environment. The formation of metal sulfides, particularly iron sulfide, is extremely important in mineral cycling and deposition.[279, 340] In saturated soils containing soluble sulfates, the activities of these microorganisms are thought to be the main factors contributing to increases in alkalinity and decreases in soluble calcium and magnesium. Abd-el-Malek and Rizk[1, 2] have proposed that sodium lactate ($2CH_3CHOHCOONa$) reacts with sodium sulfate (Na_2SO_4) to form sodium acetate (CH_3COONa), sodium bicarbonate ($NaHCO_3$), and hydrogen sulfide:

$$2CH_3CHOHCOONa + Na_2SO_4 \rightarrow 2CH_3COONa + 2NaHCO_3 + H_2S \quad (5)$$

The biochemical activity of these sulfate-reducing bacteria is also involved in the corrosion of iron under anaerobic conditions in which, in the absence of oxygen, *Desulfovibrio desulfuricans* can act as a "catalyst" in the depolarization of the corrosive action.[327] The following equations illustrate the process:

Anodic Solution of Iron

$$8H_2O \rightarrow 8H^+ + 8OH^- \quad (6)$$

$$4Fe^0 + 8H^+ \rightarrow 4Fe^{2+} + 8H \quad (7)$$

Depolarization

$$CaSO_4 + 8H \xrightarrow{D.\ desulfuricans} H_2S + 2H_2O + Ca(OH)_2 \quad (8)$$

Corrosion Products

$$Fe^{2} + H_2S \rightarrow FeS + 2H^+ \quad (9)$$

$$3Fe^{2+} + 6(OH)^- \rightarrow 3Fe(OH)_2 \quad (10)$$

In this instance sulfate replaces oxygen and its reduction is effected through the metabolic activity of *D. desulfuricans*. This process assumes that *D. desulfuricans* can use the hydrogen produced during the anodic solution of iron as the oxidizable substrate.

The crowns of large concrete sewer pipes fail when hydrogen sulfide generated in the flowing sewage is oxidized on the surface of the moist exposed upper portions of the conduits by *Thiobacillus* sp., resulting in the production of acid.[246] The acid interacts with the calcium salts in the concrete, causing structural damage.

Many microorganisms are involved in the cycling of sulfur in the environment, particularly in the transformation of sulfur species. In many instances, the impact of their activity is readily apparent, e.g., in the production of acid mine wastes, the anaerobic corrosion of iron, the failure of concrete pipe crowns, and certainly in the production of obnoxious odors. The biogenic production of hydrogen sulfide can also affect soil conditions, crop growth, and aquatic life. As little as 0.86 mg/liter of water can be toxic to trout.[215] Hydrogen sulfide can affect the taste of drinking water at levels as low as 0.05 mg/liter.[60]

Miners of gypsum, sulfur, and lead, drillers and refiners of high sulfur petroleum, sewer workers, and workers in industries where there may be biogenic hydrogen sulfide must be extremely cautious when working in confined locations. For example, in a rendering plant in Ohio, six men recently died of asphyxiation while working in a drainage sump containing biogenic hydrogen sulfide.[82]

Interest in the sulfur cycle has been growing among those involved in air quality. Sulfur compounds, particularly sulfur dioxide and hydrogen sulfide, are known to be common air contaminants. Because of the newly heightened interest, estimates of global sulfur flux have been made in an attempt to estimate biogenic and anthropogenic contributions to the sulfur cycle. These estimates only give values for the observable sulfur flux between the atmosphere and the earth's surface; the gross turnover in nature is not well understood. Table 2-1 presents estimates of annual global sulfur emissions and depositions. It is an excellent summary of the sulfur cycle.

The difference in sulfur flux between the two hemispheres is due to intensive industrial activities in the northern hemisphere (anthropogenic sulfur dioxide and sulfate) and a smaller total land mass in the southern hemisphere, which accounts for reduced biogenic emissions (natural excess sulfate), since only terrestrial areas and the littoral and estuarine

Table 2-1. Sources and deposition of atmospheric sulfur in metric tons × 10^6 of sulfur per year[a]

Origin of sulfur	Northern hemisphere		Southern hemisphere	
	Source	Deposition	Source	Deposition
Sea salt, sulfate	20	20	23	23
Diffusion of sulfur dioxide, sulfate[b]	—	19	—	5
Excess sulfate in rain[c]	—	86	—	28
Anthropogenic sulfur dioxide, sulfate	46	—	3	—
Natural excess sulfate[d]	59	—	30	—
Total	125	125	56	56

[a] Adapted from Kellogg *et al.*, 1972.[167]
[b] Diffusion to land and sea surface.
[c] Amount in excess of sea salt.
[d] The source of this excess is primarily biogenic hydrogen sulfide, which is oxidized to sulfate or sulfur dioxide in the atmosphere.

areas of the ocean are significant sources of hydrogen sulfide. It is assumed that any hydrogen sulfide produced in the open sea would be oxidized in the water column before it could reach the atmosphere. The biogenic hydrogen sulfide from land and coastal areas that reaches the atmosphere is rapidly oxidized to sulfur dioxide and sulfate. This is shown in Table 2-1 as "natural excess sulfate." An estimated 89 × 10^6 metric tons of biogenic sulfides and 0.6 × 10^6 metric tons from volcanic sources are emitted annually into the atmosphere. This amounts to approximately 47% of the total sulfur emission in the northern hemisphere. In certain areas it will contribute to the sulfur dioxide-sulfate concentration in the ambient air.

An important contribution to the sulfur cycle, which is not usually included in the classic scheme, is the flux of sulfate from sea salt. This source amounts to 20% to 40% of the total sulfur deposited, of which 2% to 10% is deposited on the land.[167, 259]

Based on the data presented in Table 2-1, the proportion of sulfur emissions from the various sources can be determined. Anthropogenic activities, primarily the combustion of fossil fuel, generate 37% of the total atmospheric sulfur source in the northern hemisphere and 27% in the southern hemisphere. Biogenic sulfur, e.g., hydrogen sulfide, accounts for the major input in both hemispheres, amounting to 49%. Since 1937 the anthropogenic contribution to total sulfur emissions has increased by ~30%; by the year 2000 these two sources should be equivalent. What impact this will have on the world sulfur cycle is unknown.

In this chapter, the sulfur cycle has been discussed primarily from the microbial point of view. These life forms are responsible for worldwide biogenic sulfur. Hydrogen sulfide appears to be a key sulfur compound in this cycle in terms of both local and global impact. The global hydrogen sulfide flux is estimated by determining the volume of sulfide required to balance mass transfer needs and is seldom based on direct measurement. Some investigators[202] feel that volatile organic sulfur compounds, particularly dimethyl sulfide, are more important in the global sulfur budget than hydrogen sulfide. There are also those[206] who claim that the contribution of dimethyl sulfide is only a small fraction of this flux. Which of the two concepts is correct remains to be seen. As discussed previously, local imbalances in the cycle produce a number of problems, including that of odors. On a global scale, biogenic hydrogen sulfide emitted to the atmosphere is converted to sulfate, which appears to be the main transport species in the flux between the earth's surface and the atmosphere. Also global, although difficult to quantify, are the "inbalance" activities involved in the sulfur cycle and their impact on mineral recycling and agriculture.

3

Absorption, Distribution, Metabolism, and Excretion of Sulfides in Animals and Humans

ABSORPTION

Hydrogen sulfide in aqueous solution has two acid dissociation constants. Dissociation of the first proton results in the formation of the hydrosulfide anion (HS^-). Dissociation of the second proton results in the formation of the sulfide anion ($S^=$). In 0.01 to 0.1 *N* solutions at 18 C, the pK_a for step 1 is 7.04, whereas the pK_a for step 2 is 11.96.[331] At the physiologic pH of 7.4, about a third of the total sulfide exists as the undissociated acid, about two-thirds as the hydrosulfide anion, and only infinitesimal amounts as the sulfide anion. Dissolved undissociated hydrogen sulfide maintains a state of dynamic equilibrium with gaseous hydrogen sulfide at the air-water interface.

These properties are highly relevant to the biologic effects of sulfide. The undissociated acid is a more potent inhibitor of cytochrome oxidase than is the anionic form. Insofar as the systemic effects of sulfide are due to the inhibition of cytochrome oxidase, systemic acidosis would intensify them. On the other hand, the anionic moiety complexes with methemoglobin when the latter is therapeutically induced (cf. Chapter 4). In accord with the principles of nonionic diffusion, it is likely that undissociated hydrogen sulfide crosses biologic membranes more rapidly than the charged anionic species. This supposition is supported by data collected by Beerman[25] on the effects of hydrogen sulfide on various protozoan species. At least one interpretation of his results is that sulfide penetrates into cells more rapidly as the un-ionized moiety. Similarly, the absorption of sulfide from the peritoneal cavity of mice appeared to be accelerated by an acidic environment and delayed by an alkaline one.[286] Because sodium sulfide is promptly and completely hydrolyzed in aqueous solutions, these considerations apply to solutions of the salt as well as to the acid.

With very few exceptions, accidental sulfide poisonings have resulted from respiratory exposure to the gas (cf. Chapter 5), but the question of whether or not hydrogen sulfide can be absorbed through the skin in amounts that are toxicologically significant was addressed by very early investigators. In 1803, Chaussier was able to produce death in animals when he exposed their bodies to the gas while they were breathing fresh air.[220] Even though some investigators, such as Yant,[338] deny that hydrogen sulfide is absorbed through intact skin, most report that systemic effects of sulfide and evidence for its pulmonary excretion are detected after cutaneous exposure of animals.[20, 184, 250] However, quantitative data are lacking, and death of animals may occur only after extensive exposure to large areas of the skin to pure hydrogen sulfide.[330] Industrial experience suggests that percutaneous absorption must be many times less efficient than pulmonary absorption.

Solutions of hydrogen sulfide administered orally or as enemas produce prompt evidence of systemic absorption; again, quantitative data are lacking. Solutions of sodium sulfide are highly alkaline and corrosive. Presumably, their ingestion would result in local effects like those of lye as well as systemic effects due to sulfide. Hydrogen sulfide that has been generated by the intestinal microflora is, at least in part, systemically absorbed and detoxified.[85]

DISTRIBUTION AND EXCRETION

Confusing and partly contradictory findings have been reported about the distribution of sulfide in the body. At least two reports suggest that very little sulfide is found in the brains of animals given sodium sulfide by mouth.[70, 91] Experimental evidence, however, strongly suggests that sulfide causes death by a central nervous system action (cf. Chapter 4). Systemically administered sulfide appears to be concentrated in the liver with smaller proportions in the kidneys and the lungs.[70, 175] After the sodium salt was given to rats by mouth, 50% of the ^{35}S label appeared in the urine as sulfate within 24 hr. When administered intraperitoneally, 90% of the label was recovered in the urine and feces in 6 days.[91]

Unfortunately, no reliable estimates appear to be available on the quantitative importance of pulmonary excretion of hydrogen sulfide. Although many workers have noted the presence of hydrogen sulfide in expired air after its administration, a systematic investigation of that phenomenon is indicated, i.e., how does the amount excreted via the lungs vary as a function of the time after administration, as a function of the route of administration, and as a function of the species? If a significant fraction of the total dose is eliminated by the pulmonary route, it could account for the observed value of artificial respiration, particularly in the presence of apnea. Moreover, forced hyperventilation might constitute an important therapeutic procedure for hastening sulfide excretion.

METABOLISM

A sulfide oxidase system, which exists in rat livers and kidneys, catalyzes the oxidation of sulfide to thiosulfate. This system may contain both a heat-stable and a heat-labile component, and it is most closely associated with the mitochondrial fraction of rat liver cells.[23] Some evidence suggests that sulfite may be an intermediate in the reaction, which could involve scission of a disulfide bond on a protein to form a thiosulfonate.[21] The activity of the rat liver sulfide-oxidizing system, however, could be mimicked by adding iron in physiologic concentrations to albumin. Ferritin was found to be even more active in oxidizing sulfide to thiosulfate than the rat liver system.[22]

Sörbo[292] believes that the bulk of sulfide oxidation *in vivo* proceeds nonenzymatically. He has shown that a variety of iron-containing compounds, including hemin, can catalyze the reaction. Hemoglobin, myoglobin, catalase, and cytochrome *c*, however, were among those compounds that were inactive. He has also suggested that the conversion of sulfides to polysulfides allows rhodanese to act on the latter to generate thiocyanate.[293] Autoxidation of hydrogen sulfide results in the formation of hydrogen peroxide, which, Bittersohl[37] suggests, may be responsible for some of the toxic effects of sulfide in the central nervous system. However, convincing *in vivo* evidence is lacking.

ENDOGENOUS HYDROGEN SULFIDE PRODUCTION

Huovinen and Gustafsson[143] found only very low levels of incorporation of [^{35}S]sulfate and [^{35}S]sulfite into cysteine and methionine in ordinary rats and none at all in germ-free animals. On the other hand, incorporation of [^{35}S]sulfide into cysteine occurred extensively in both conventional and germ-free rats, but incorporation into methionine was limited. Thus, rat tissues do not appear to be able to reduce sulfate or sulfite to sulfide, although the intestinal microflora have a limited capacity for carrying out such reductions.

In 1927, Denis and Reed,[85] after a survey of earlier literature, concluded that hydrogen sulfide is "probably constantly present in the large intestines as a result of the bacterial decomposition of proteins." They also believed that the rate of hydrogen sulfide production would be very difficult to measure because the compound is so rapidly absorbed from the lumen. Coupling the foregoing with the known high toxicity of sulfide also suggests the presence of a highly efficient systemic detoxication mechanism. In 1937–1938, Andrews[13] documented the production of hydrogen sulfide, from different sulfur-containing substrates, by minced mucosa from the small intestines of dogs. A variety of bacteria generate hydrogen sulfide (cf. Chapter 2); several yeasts produce significant amounts.[256]

From analyses of one patient with a lactose malabsorption syndrome, Levitt *et al.*[190] concluded that there were only five components of intestinal gas, but they observed rather remarkable differences in its composition after various test diets were administered: nitrogen, 18% to 91%; oxygen, 0.04% to 4.1%; hydrogen, 0.02% to 41%; carbon dioxide, 5% to 38%; and methane, 0.001% to 28%. Apparently hydrogen sulfide was not detected. According to Danhof,[81] the normal composition of flatus in humans is: nitrogen, 55%; methane, quite variable, but 14% to 15% in one subject; oxygen, 12% to 14%; carbon dioxide, 12%; and hydrogen, 4% to 5%. In contrast, Saltman and Sieker,[264] after a survey of literature values, reported the composition to be: nitrogen, 70% to 86%; carbon dioxide, 6% to 12%; oxygen, 0% to 12%; hydrogen, 1% to 10%; methane, 0.1% to 2%; and hydrogen sulfide, 0% to 10%. Obviously, there is little unanimity on this subject and further studies are warranted.

Among the factors that may influence the concentration of hydrogen sulfide in flatus are the degree, duration, and level of intestinal obstruction. Cantor and Weiler[61] described the discoloration of intestinal decompression tubes by hydrogen sulfide in the bowels of patients. They indicated that hydrogen sulfide may constitute as much as 12% of intestinal gases in severe cases. Although additional work is desirable, these reports suggest that there may be pathophysiologic conditions under which hydrogen sulfide accumulates in the human intestines. Such relatively modern reports lend some credence to older literature about an association between constipation and the production of abnormal blood pigments in humans (cf. Chapter 4).

Both hydrogen sulfide and methyl mercaptan have been associated with oral malodor. As measured in 10 subjects over 6 to 10 days, the mean values per subject ranged from 65 to 698 ppb of hydrogen sulfide and from 10 to 188 ppb of methyl mercaptan.[38]

4

Effects on Animals

As indicated in the article by Mitchell and Davenport,[220] which is reprinted in Appendix II of this text, hydrogen sulfide poisoning attracted the attention of some of the best biologic experimentalists of the nineteenth century. The excitation and respiratory stimulation produced by inhalation of the gas or by injection of solutions of hydrogen or sodium sulfide were characterized in the mid-1800's. The high toxicity of hydrogen sulfide, its lethal propensity to produce respiratory arrest (apnea), sometimes with attendant convulsions, and the value of artificial respirations in averting death were also known at that time. Several workers noted that part of a parenteral dose was excreted via the lungs.

The pervasive but erroneous hypothesis that sulfide was a blood poison like carbon monoxide or sodium nitrite was firmly entrenched even though workers were unable to demonstrate a significant accumulation of "sulfhemoglobin" in the blood of poisoned animals.

Both early experiments with animals and accounts of accidental human exposures indicated that hydrogen sulfide was a significant irritant after prolonged exposure to low concentrations. (The irritant properties of airborne chemicals are not necessarily related to their odor.[9] See also Chapter 8.) Inflammation of the eyes and mucous membranes was observed in sewer and tunnel workers who had been exposed chronically to hydrogen sulfide. In its extreme form, this irritant activity has led to the development of fatal pulmonary edema, as distinct from the respiratory arrest without pulmonary involvement seen after exposure to higher concentrations of the gas.

ACUTE POISONING BY SULFIDE

Respiratory and Circulatory Effects of Sulfide

In experiments with dogs, Haggard *et al.*[129] noted marked differences in the effects of hydrogen sulfide with only small changes in the concentration in the inspired air. Exposure of dogs to what was considered to be a minimal lethal concentration (0.05% by volume in air) resulted in a slight progressive depression of the rate and depth of respirations. After many hours of exposure, animals died from pulmonary edema. When the concentration of hydrogen sulfide was doubled (to 0.1% by volume), death resulted in 15 to 20 min. In this case the respiration was stimulated almost immediately. The stimulation progressed to a violent hyperpnea, which was followed by death in apnea. At 0.3% by volume in the inspired air, respiratory arrest occurred after a few violent gasps.

Except for pulmonary edema, the same effects on respiration could be elicited by the intravenous administration of sodium sulfide. Doses of 2 to 4 mg/kg resulted in immediate hyperpnea followed by variable periods of apnea for which artificial respiration was instituted. Haggard *et al.*[129] presented evidence that the respiratory stimulant effects of sulfide were abolished by vagotomy. After vagotomy (at an unspecified anatomic locus), sulfide administration resulted only in respiratory depression. This alleged effect of vagotomy, however, was to remain the controversial aspect of their work.

A fuller explanation of the respiratory stimulant activity of sulfide had to await the discovery of the chemoreceptor function of the carotid body and the reflex effects that are secondary to the activation of these receptors. Elucidation of the function of the carotid body emerged from experiments conducted by J. Heymans and his son, C. Heymans,[138] in Belgium in the early 1930's. In 1938 the younger Heymans received the Nobel prize for these studies. He found that cyanide and sulfide were among the agents that activate the carotid chemoreceptors. The identical effects of these two agents, as mediated through the reflexes initiated by the carotid body, consitute part of the evidence that cyanide and sulfide have similar toxic mechanisms of action.

Heymans *et al.*[136] showed that the injection of a small dose of sodium sulfide into the common carotid artery of dogs (Figure 4-1 shows the very similar anatomy of the cat) resulted in an immediate and powerful respiratory excitation. After denervation of the carotid sinus by section of the sinus nerve, the injection of even larger doses of sodium sulfide had no immediate effect on respiration, and the late effect tended to be that of depression of the respiration. Similarly, the injection of large doses of sulfide into the internal carotid or vertebral arteries also failed to elicit respiratory stimulation, presumably because sulfide injected at those sites would reach the carotid body only after dilution in the general circulation. Thus, the findings of Heymans *et al.*[136] are somewhat parallel to those of Haggard *et al.*,[129] except that Heymans *et al.* sectioned the sinus nerve instead of the vagus nerves to block the effects of sulfide.

Heymans *et al.*[136] also used cross-perfusion techniques to show that sulfide acted on the carotid chemoreceptors. In these experiments the isolated carotid sinuses of a recipient dog received their entire blood supply from a donor dog. Sulfide given systematically to the recipient dog had no effect on respiration, whereas sulfide given systemically to the donor dog provoked the typical response in the recipient dog.

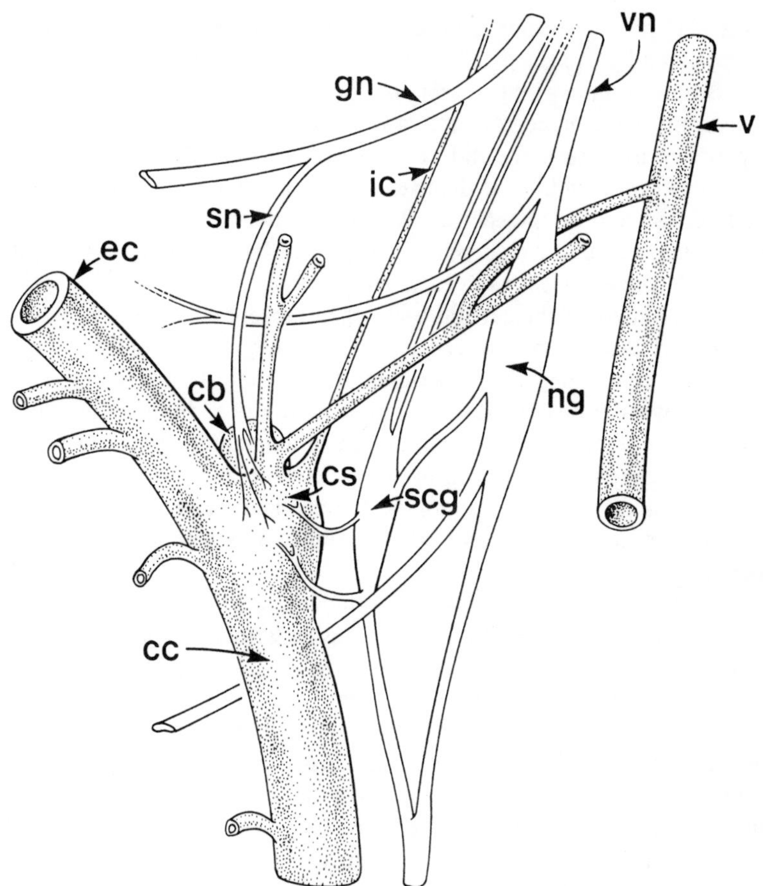

Figure 4-1. The left carotid bifurcation and associated nerves of the cat from the ventral side. Abbreviations: cb, carotid body; cc, common carotid artery; cs, carotid sinus; ec, external carotid artery; gn, glossopharyngeal nerve; ic, internal carotoid artery; ng, nodose ganglion; scg, superior cervical ganglion; sn, sinus nerve; v, vertebral artery; vn, vagus nerve. (Redrawn and derived from Adams, 1958.[5])

Chemoreceptor activity associated with the aortic bodies was also known to Heymans *et al.*,[136] but this function appeared to play little or no role in the physiologic responses to cyanide or sulfide, i.e., denervation of the carotid sinus alone usually blocked the response to cyanide or sulfide completely.

When the innervation of the carotid sinus was intact, the respiratory response to cyanide or sulfide was accompanied by a vasomotor reaction. Doses as low as 1 μg/kg of sodium sulfide injected into the common carotid artery provoked a fleeting rise in the systemic blood pressure.[137]

Since this pressor response occurred also in curarized animals, it could not be secondary to the mechanical events of the hyperpnea.

Bradycardia was also seen on occasion. Therefore, the response to carotid chemoreceptor activation by cyanide or sulfide includes hyperpnea, hypertension, and perhaps bradycardia. Bradycardia was the most inconstant finding. It depended, among other factors, on the site of the injection and on the species. Whether or not this slowing of the heart is mediated through concurrent activation of baroreceptors by sulfide or cyanide has never been resolved satisfactorily.[138]

Stimulation of the carotid body chemoreceptors by sulfide was confirmed in a brief report by Owen and Gesell,[242] which contained almost no details concerning the experimentation. They did, however, include the controversial observation that the injection of sodium sulfide into the fourth brain ventricle also results in an "immediate, well-sustained, and marked augmentation of ventilation." Winder and Winder[334] failed to confirm this direct central action to sulfide. They suggested that it might have been an artifact related to the alkalinity of decomposed solutions of sodium sulfide.

In all other respects, however, Winder and Winder[334] confirmed and extended the findings of the previous investigators. In their intact dogs, which had been anesthetized with morphine and urethane, an abrupt, intense augmentation of pulmonary ventilation, particularly in the depth of respiration, followed the injection into the common carotid of sodium sulfide at doses as low as 0.6 μg/kg. This response was identical to that elicited by comparable doses of cyanide, and it was eliminated by carotid sinus denervation. Denervation of the sinus or any maneuver by which the injected bolus of sulfide was caused to bypass the carotid sinus resulted only in late respiratory depression, which was presumed to be mediated directly through the brain stem nuclei concerned with the integration of respiratory movements.

In an early attempt to localize and characterize the brain stem nuclei that control respiratory movements, Stella[295] decerebrated cats below the "pneumotaxic center"† by transection of the pons a few millimeters caudal to its upper border. The "apneustic center," which was presumed to be responsible for inspiratory activity, remained intact. Cyanide and sulfide, which were injected into the common carotid, produced their characteristic respiratory stimulation (and also "apneustic stimulation," according to Stella). Denervation of the carotid sinus blocked the response as usual. Thus, both afferent and efferent limbs of the carotid reflex were localized to a particular level of the brain stem. That area included the fourth ventricle.

---

† Controversy surrounds the function of the pneumotaxic center, but it is generally believed to coordinate inspiratory and expiratory activity.

A series of potentially interesting observations on the carotid reflex response to sulfide were made by Koppanyi and Linegar in 1942;[173] but, unhappily, these were published only in abstract form. In mammals the bradycardia that resulted from sulfide administered intravenously in doses of 0.5 to 10 mg/kg was eliminated by vagotomy; therefore, it presumably represents part of the reflex response elicited by sulfide. The investigators insisted, however, that the pressor response was due to a peripheral effect of sulfide because it was not blocked by nicotine. Moreover, it persisted after the adrenal veins were clamped. Of even greater interest was their assertion that sectioning of the vagus nerves in the neck did not abolish the carotid reflex hyperpnea after exposure to sulfide, whereas sectioning of the vagus nerves above the nodose ganglion did block the response. These observations do not appear to have been reconciled with those of Haggard *et al.*,[129] Heymans *et al.*,[136] and later workers.

Medvedev[216] reported that sulfide influences transmission through the superior cervical sympathetic ganglia of cats. Low concentrations (0.2 to 0.5 mg/ml in the perfusate) had excitatory effects that were blocked by atropine, whereas higher concentrations (0.6 to 1.0 mg/ml) had inhibiting effects that were reversed by cocaine.

After this brief flurry of activity, sulfide was employed infrequently as a tool for the study of respiration. Some additional work, however, was performed by Russian scientists. A good English-language summary of this work can be found in the monograph by Anichkov and Belen'kii.[14] Russian workers were convinced that part of the reflex response to carotid chemoreceptor activation did in fact involve adrenal discharge or a generalized sympathetic discharge. For example, after exposure to cyanide or sulfide, contraction of the splenic capsule, an increase in the number of circulating red cells, and transient hyperglycemia were observed in cats. Increased circulating concentrations of 17-hydroxycorticosteroids were held responsible for the eosinopenia.

The most recent and extensive study of the respiratory and circulatory effects of sulfide was conducted by the late C. Lovatt Evans.[103] He also confirmed that doses in the range of 20 μmol/kg (1 mg/kg), when rapidly injected into cats intravenously, provoked a marked hyperpnea that was often followed by permanent respiratory arrest. The hyperpnea was usually blocked by local anesthesia of the carotid sinus region, although in one case, when sulfide was injected into the ascending aorta, hyperpnea still occurred.

Evans,[103] like Winder and Winder[334] before him, was unable to induce hyperpnea by injecting sulfides into the fourth brain ventricle. Only respiratory depression or arrest was elicited by that route, but a large and prolonged rise in the arterial blood pressure followed doses as low as 1.5 μmol/kg (0.05 mg/kg). This suggests a direct central pressor action of sulfide, whereas the results obtained by Koppanyi and Linegar[173] suggested a peripheral pressor action. When sulfides were injected intravenously, functional denervation of the carotid sinus abolished the transient pressor response due to sulfide. Only the late depressor effect was observed. Evans[103] agreed with Koppanyi and Linegar that the bradycardia seen in cats after exposure to sulfide was eliminated by vagotomy. Various other arrhythmias, disorders of conduction, and impairment of ventricular repolarization in both humans (Chapter 5) and animals[176] have also been reported. Perhaps these represent direct toxic actions secondary to an impairment of oxidative metabolism. Histopathologic changes in heart muscle[176] and brain[203] are similar to those observed after hypoxia due to oxygen lack or carbon monoxide poisoning.

Obviously, not all of the observations listed above can be generally true; however, it is certain that the respiratory effects of sulfide are critical to the acute intoxication syndrome. Like cyanide, sulfide stimulates carotid body chemoreceptors to produce a dramatic hyperpnea. The ultimate cause of death, however, is respiratory arrest, which is attributed to the direct depressant effects of sulfide at the level of the brain stem. From time to time it has been suggested that effects of cyanide on respiration are mediated through mechanisms in addition to carotid body reflexes.[46, 189] Such possibilities should be considered when planning future experimentation with sulfide.

The literature concerning the less significant cardiovascular effects of sulfide is even more confusing. The simplest explanation that fits many (but certainly not all) observations is that the transient rise in blood pressure is due to a sympathetic discharge that is secondary to a carotid reflex. Although also a reflex, the bradycardia is probably not secondary to the increase in blood pressure. Additional experimentation is needed to confirm or refute these speculations.

Effects of Sulfide on the Blood

Sulfhemoglobin The term "sulfhemoglobin" was coined by Hoppe-Seyler[220] for the spectrally altered pigment that was generated when pure hydrogen sulfide was bubbled through blood *in vitro*. Confusion in the literature about this isolated observation persists even today. For example, the early medical literature contains many reported examples of alleged "sulfhemoglobinemia" of obscure etiology that occurred in human patients on a more or less chronic basis (e.g., Wallis, 1913–1914).[329] The source of sulfide was believed to be the intestine, where it was suspected that hydrogen sulfide was generated as a result of biochemical reactions carried out by the microflora (see Chapter 3).

These early reports of cases of sulfhemoglobinemia cannot be regarded as reliable. At that time the spectroscopic instrumentation was crude and there was insufficient awareness of other pathophysiologic states that could confuse the diagnosis. A specific spectrophotometric procedure for the quantitative determination of methemoglobin was not published until 1938.[104] Before then, the diagnosis of sulfhemoglobinemia was probably sometimes confused with that of methemoglobinemia. The failure of the patient's history to reveal exposures to drugs or chemicals known to be responsible for methemoglobinemia was not necessarily helpful in the differential diagnosis; the methemoglobinemia could have been due to the inheritance of an abnormal hemoglobin (e.g., one of the hemoglobins M) or to a deficiency in the red cell enzyme, methemoglobin reductase.

The congenital methemoglobinemias and other hemoglobinopathies are rather recent discoveries. In the first half of the twentieth century many cases were, in all likelihood, lumped together under the term "enterogenous cyanosis." It was presumed that intestinal infections led to the elaboration of chemicals that generated methemoglobin systemically or facilitated the formation of sulfhemoglobin from endogenous hydrogen sulfide (see Chapter 3). In 1948 Finch[110] reported an association between constipation and sulfhemoglobinemia. Although it is tempting to speculate about this fascinating older literature, such an exercise is unlikely to be productive because of the uncertainties mentioned above. Even the term "enterogenous cyanosis" has largely disappeared from modern reference works.

The procedure for the determination of methemoglobin and sulfhemoglobin reported in 1938 by Evelyn and Malloy[104] remains the most widely used technique even today. For better or for worse, the compounds are defined in terms of this procedure. There are no objections to the definition of methemoglobin as an abnormal blood pigment, with an absorption maximum at 635 nm, that is abolished on the addition of cyanide. Less satisfactory is the definition of sulfhemoglobin as an abnormal blood pigment, with an absorption maximum at about 620 nm, that is unchanged by the addition of cyanide. Most notably, this definition fails to include a statement about the etiology of the pigment formation. Moreover, such abnormal blood pigments are loosely referred to as "sulfhemoglobins," whether or not hydrogen sulfide was known to have been involved in their formation.

Many investigators noted that the blood of patients receiving "oxidant" drugs such as phenacetin often contained small amounts of pigments that fit the general definition of sulfhemoglobin. Indeed, the patients selected by Evelyn and Malloy[104] for a definitive test of their procedure were receiving sulfanilamide, which is notorious as a methemoglobin former but has never been established as a source of hydrogen sulfide. The abnormal absorption maximum at 620 nm remaining in their blood after the addition of cyanide was assumed to be identical to that exhibited by solutions of blood that had been exposed to hydrogen sulfide. In retrospect it seems probable that that assumption was incorrect. The application of the term "sulfhemoglobin" to two distinctly different phenomena has resulted in an almost hopeless confusion in the literature.

Unhappily, this confusion cannot be resolved even today because the chemistry of neither "sulfhemoglobin" has yet been precisely defined. There are methods for preparing sulfhemoglobin in high yield from sulfide and oxyhemoglobin, but the product has not yet been characterized in rigorous detail.[153] Even so, there are many indications that that pigment bears little or no relationship to the more widely known entity generated during oxidative hemolysis by phenylhydrazine and related redox compounds, both in normal red cells and, more spectacularly, in red cells deficient in glucose-6-phosphate dehydrogenase. In that reaction the sequence of events is irreversible and appears to involve the formation of methemoglobin, "sulfhemoglobin," and Heinz bodies, followed by intravascular hemolysis. Again, the "sulfhemoglobin" generated in this reaction has not been defined precisely, but presumably it involves the formation of mixed disulfides between glutathione and globin sulfhydryl groups.[152] Since the reaction can be elicited *in vitro* in simple systems, it hardly seems likely that sulfide *per se* is an obligatory participant. This entity is probably a mixture of abnormal and partially denatured pigments, and it needs another name—perhaps pseudosulfhemoglobin. In this report, sulfhemoglobin designates pigments known to have been generated in the presence of sulfide.

Experiments in vivo Because the effects of hydrogen sulfide on blood *in vivo* are in some ways less controversial than their effects *in vitro*, they are given precedence in this discussion. Despite the analytic crudities and the vagaries of the definitions indicated above, early reports on the effects of hydrogen sulfide on the blood of humans or animals dying from acute poisoning are clear and unambiguous. Neither sulfhemoglobin nor any other abnormal blood pigments were found. [127, 128, 329, 338] Thus, whatever sulfhemoglobin may be, it is not generated *in vivo* in life-threatening concentrations in acute sulfide poisoning.

Sulfide also affects blood by decreasing the oxygen content.[103, 127] This presumably results in the accumulation of deoxyhemoglobin. Despite statements to the contrary in textbooks, it is highly improbable that the oxygen transport capability of the blood could be compromised significantly by such a mechanism. As long as the respiration and circulation continue, deoxyhemoglobin would be continuously reoxygenated in the lungs.

Two more recent observations confirm that sulfide does not interrupt oxygen transport by the blood in the acute intoxication syndrome.[273, 287] First, the concomitant induction of methemoglobinemia actually protects animals against death. If sulfide compromised oxygen transport, methemolgobinemia would exacerbate the intoxication. Second, oxygen (100% at 1 atm) has no prophylactic or therapeutic effects in terms of the mortality of mice injected with sulfide compared with groups of animals treated similarly but maintained under air at 1 atm.[289]

Although it cannot play a significant role in acute intoxication, sulfhemoglobin does appear as a post-mortem change in the blood of individuals who died as a result of exposure to hydrogen sulfide.[127, 128] Presumably, this phenomenon is partially responsible for the peculiar pigmentations noted in various organs during autopsy (see Chapter 5). At least one report indicates that concentrations of 15% to 20% sulfhemoglobin could be generated in rabbits by applying ammonium hydrogen sulfide to their clipped skin for periods of 2 to 24 hr. When death resulted in less than 12 min, no sulfhemoglobin was detectable in the blood.[184] These observations need confirmation. It would be of further interest to explore the possible role of sulfhemoglobin in relation to subacute or chronic exposures to hydrogen sulfide.

Experiments in vitro Considering the fact that sulfhemoglobin has never been generated *in vivo* in concentrations high enough to cause toxic signs, the continued interest in this pigment over the years is remarkable. It is not possible or even desirable to review here all the experimentation on this subject. Yet, because the significance and even the chemical nature of sulfhemoglobin remain in doubt, a sampling of that literature is clearly in order.

In 1907, Clarke and Hurtley[68] were apparently the first to discover that carbon monoxide can complex with sulfhemoglobin. A new entity, referred to as carboxysulfhemoglobin, could be obtained by gassing sulfhemoglobin with carbon monoxide or carboxyhemoglobin with hydrogen sulfide. They also confirmed earlier work that blood deoxygenated with carbon dioxide became resistant to the spectral changes induced by sulfide. In contrast, a number of "reducing" substances, including phenylhydrazine, appeared to sensitize the blood to sulfide. Unlike inorganic sulfide, ethyl mercaptan had no effect on blood. Like many workers both before and after them, Clarke and Hurtley were troubled by the apparent instability of the product and its obvious impurity.

In 1935–1936, the first detailed analysis of the complete visible absorption spectrum of sulfhemoglobin was made by Drabkin and Austin.[88] They concluded that all previous spectral studies had been conducted on mixtures of sulfhemoglobin and deoxyhemoglobin. An absorption spectrum of "pure" sulfhemoglobin was derived from these mixtures by correcting for the absorbance of deoxyhemoglobin.

Deoxyhemoglobin is a product of the reaction between sulfide and oxyhemoglobin, but deoxyhemoglobin itself could not be converted into sulfhemoglobin until some oxygen was admitted into the system.

Although any soluble inorganic sulfide could function in the reaction under the appropriate conditions, Michel[217] reported that hydrogen peroxide was essential for sulfhemoglobin formation. The heme moiety of Michel's product was not irreversibly changed. The solubility, resistance to alkali, molecular weight, and electrophoretic mobility of his sulfhemoglobin were the same as for hemoglobin.

Morell *et al.*[225] described experiments on sulfmyoglobin, a pigment prepared from myoglobin that is presumably analogous to sulfhemoglobin. Indeed, sulfmyoglobin may be easier to prepare and is perhaps more stable than sulfhemoglobin. Sulfmyoglobin contains one more sulfur atom per molecule than myoglobin, a finding analogous to that of Michel[217] for sulfhemoglobin, which contained one more sulfur atom per heme group than hemoglobin. Morell *et al.*[225] suggested that although the sulfur atom is bound in a manner that affects the visible absorption spectrum of the heme group, it does not appear to be attached to the iron *per se*. Most likely the sulfur adds across a pyrrole double bond in the porphyrin. Similar reaction sequences have been proposed for the generation of sulfmyoglobin and sulfhemoglobin, both of which require the participation of hydrogen sulfide and hydrogen peroxide.[225, 234] An alternative scheme proposed by Frendo[112] suggests that sulfhemoglobin can be generated in the presence of elemental sulfur if glutathione or other mercaptans are present to convert sulfur to hydrogen sulfide.

The reversible nature of sulfhemoglobin was again pointed out in 1970 by Johnson.[153] Like the early workers, he noted that the rate at which sulfhemoglobin formed from oxyhemoglobin and sulfide depended in a complex way on the oxygen tension. When excess oxygen was removed by vacuum, much of the sulfhemoglobin originally formed was slowly reconverted to deoxyhemoglobin. The original decrease in the oxygen tension was insufficient in and of itself to result in significant accumulation of deoxyhemoglobin. Johnson described a number of complexes of ferro- and ferri-sulfhemoglobin, including the bizarre carboxycyano-ferro complex, which apparently requires six coordination positions, thus excluding the globin histidine that normally occupies one of those positions.

The apparently reversible nature of sulfhemoglobin, which is generated in the presence of sulfide, distinguishes it from pseudosulfhemoglobin, which is generated by redox compounds in the absence of sulfide. Pseudosulfhemoglobin is thought to represent irreversible alterations in the pigment that persist for the life of the red cell.[156]

In summary, the formation of sulfhemoglobin from oxyhemoglobin and sulfide is a complex process that may involve a delicate balance between reductive and oxidative reactions. It has been difficult to study the reaction because of the turbidity due to the formation of elemental sulfur and denatured hemoglobin products. The problem is further confounded by the apparently unstable nature of sulfhemoglobin itself, which tends to regenerate hemoglobin. This pigment probably bears no relationship to the so-called "sulfhemoglobin" that is generated in normal red cells during chemically induced oxidative denaturation of hemoglobin where a soluble inorganic sulfide is not known to play a role.[229] Neither phenomenon is significantly involved in sulfide poisoning *in vivo*.

Effects of Sulfide on Enzymes

In view of the recognized high toxicity of sulfide, it is remarkable that there are so few reports of its effects on enzyme systems. Biochemists must have been aware of the toxicologic similarities between sulfide and cyanide because the two have almost always been compared in studies in cyanide. For example, hydrogen sulfide was said to be a more potent inhibitor of horseradish peroxidase than hydrogen cyanide.[333] Both cyanide and sulfide strongly inhibited potato "polyphenol oxidase,"[164] which is now known to be a copper-containing enzyme similar to tyrosinase.

Both cyanide and sulfide are weak inhibitors of uricase, but in some instances there are differences between these two poisons. For example, cyanide failed to inhibit amine oxidase whereas sulfide was a potent and irreversible inhibitor.[166] Another difference is the sulfide inhibition of red cell gluthathione peroxidase.[69] Succinic dehydrogenase was inhibited by sulfide, but only in the presence of oxygen. This inhibition was reversible, and it could be completely antagonized by cysteine.[35] These observations suggest that sulfide inhibition of some enzymes may involve scission of a critical disulfide bond, a mechanism that is unlikely to occur with cyanide. Under physiologic conditions sulfide has a much greater propensity to interact with disulfide bonds than does cyanide.

With the discovery that cyanide and sulfide had similar inhibitory effects on catalase,[297] it was suspected that these agents might have some special predilection for the heme-type enzymes. Azide is a third agent with a high specificity for catalase. It may also have some toxicologic properties in common with sulfide and cyanide.[42]

It was long thought that the key biochemical lesion induced by cyanide and sulfide was an inhibition of the essential respiratory enzymes. In the literature cyanide and sulfide were referred to as inhibitors of cytochrome oxidase long before a modern biochemically acceptable demonstration of such an effect was published.[164] The general acceptance of this "fact" persisted for many years[282] before a rigorous proof of it in isolated preparations finally appeared. Indeed, only in 1966 was it suggested[66] that, as an experimental tool for the terminal inhibition of respiratory chain electron transfer, sulfide might have certain advantages over the classical inhibitors, cyanide and azide.

The effects of sulfide on submitochondrial particles containing cytochrome aa_3 have not been studied in detail.[235, 283] It is clear that sulfide, like cyanide, is a slow binding inhibitor with a very high affinity for the enyzme. In fact, sulfide is a more potent inhibitor than cyanide, but azide is a very weak inhibitor. The apparent dissociation constant for the cytochrome sulfide complex is less than 0.1 μM in sulfide. Unlike cyanide, the binding of sulfide appears to be independent of the redox state of the oxidase, but the undissociated species in each case (hydrogen cyanide, hydrogen sulfide, and hydrogen azide) is more active as an inhibitor than the respective anionic species. Although it is unlikely that azide produces death by inhibition of cytochrome oxidase, such a mechanism is highly probable in acute sulfide poisoning.

In conclusion, a toxicologic similarity between sulfide and cyanide is paralleled by a recently well established similarity in their effects on a key enzyme system, the respiratory electron transport chain. Both are highly specific inhibitors of cytochrome aa_3. It is generally believed that all the pharmacologic effects of cyanide are explicable in terms of that single mechanism. The same may be true of sulfide, but sulfide has a wider spectrum of inhibitory effects on enzymes *in vitro* than does cyanide.

Antidotes to Sulfide Poisoning

Sulfmethemoglobin The existence of a complex between methemoglobin and sulfide, which is distinctly different from "sulfhemoglobin," was first recognized by Keilin in 1933.[165] This complex eventually proved to be analogous to cyanmethemoglobin, in that the sulfide was reversibly bound to the ferric heme iron. The visible absorption spectrum of this complex was distinctly different from that of either methemoglobin or sulfhemoglobin. Its formation required one molecule of hydrogen sulfide for each iron atom in the hemoglobin tetramer. It could be generated in the complete absence of oxygen. Removal of the sulfide, reduction of the methemoglobin, and aeration resulted in the regeneration of oxyhemoglobin. To distinguish this complex from sulfhemoglobin, the term sulfmethemoglobin was coined.[287]

Drabkin and Austin[88] confirmed some of the above distinctions between sulfhemoglobin and sulfmethemoglobin. They concurred with the analogy between sulfmethemoglobin and cyanmethemoglobin. When the sulfide was removed from the former and cyanide was added, the latter was generated. Coryell *et al.*[78] studied the magnetic susceptibility of the complex. They were the first to point out that, in the pH range of 5 to 7, sulfmethemoglobin apparently undergoes autoreduction to deoxyhemoglobin. When the mixture was aerated, oxyhemoglobin was formed.

This decomposition reaction was described as "rapid," although the half-life of the complex at 25 C was slightly over 2 hr.

In 1939, Coryell,[77] using Keilin's data,[165] estimated the dissociation constant as 5.3×10^{-6} M, under the assumption that only the hydrosulfide anion (HS^-) was bound to the heme iron. Any possible effect of the autoreduction reaction on the results was neglected because its rate was considered to be too slow to play a significant role under the conditions of the experiment. Thus, Keilin's[165] failure to note the decomposition of sulfmethemoglobin may be ascribed to the slowness of that reaction relative to the very rapid rate of association and dissociation of the hydrosulfide anion with ferric heme groups. Using Scheler's data[272] and correcting for undissociated hydrogen sulfide in solution, Smith and Gosselin[288] estimated the dissociation constant of the complex as 6.6×10^{-6} M; they also contributed their own estimates, which ranged from 7.5 to 20×10^{-6} M.

Despite the many clear differences between sulfhemoglobin and sulfmethemoglobin, there is at least one remarkable similarity. Both are unstable and tend to decompose to hemoglobin. This suggests that the two pigments are related in a way that has not yet been clarified. Figure 4-2 presents two simple hypothetical schemes for such an interrelationship. Scheme 1 is believed to satisfy generally all the observations described above. Scheme 2 involves an as yet unreported conversion of sulfmethemoglobin to sulfhemoglobin.

Just as one can draw analogies between sulfmyoglobin and sulfhemoglobin, a similar relationship presumably exists between sulfmetmyoglobin and sulfmethemoglobin. Indeed, the binding kinetics of hydrogen sulfide to sperm whale metmyoglobin have been examined in great detail by Goldsack *et al.*[121] As indicated above, studies with the less complex myoglobin have progressed well beyond any now available with hemoglobin, and they suggest that even the general schemes in Figure 4-2 are far too simplistic. A clear review and statement of the current status of the possible interactions of sulfide with myoglobin can be found in Peisach.[249] A ferrous sulfmyoglobin of greater than 90% purity can be prepared by the sequential reaction of hydrogen peroxide and hydrogen sulfide with *ferric* myoglobin. In this bright green pigment the sulfur is bound to the prosthetic group rather than the protein although it is not attached to the heme iron. Sulfmyoglobin reversibly binds carbon monoxide and oxygen but with less affinity than myoglobin. Sulfmyoglobin can be oxidized by a single electron transfer to sulfmetmyoglobin, which can also bind reversibly ionic ligands such as cyanide and sulfide. Thus, Peisach believes there could be a species in which sulfide is bound both to the heme iron and to some other part of the porphyrin ring. For want of better nomenclature, such a species has been designated in Figure 4-3 as sulfmetsulfmyoglobin. Even more complex species have been described above for sulfhemoglobin.

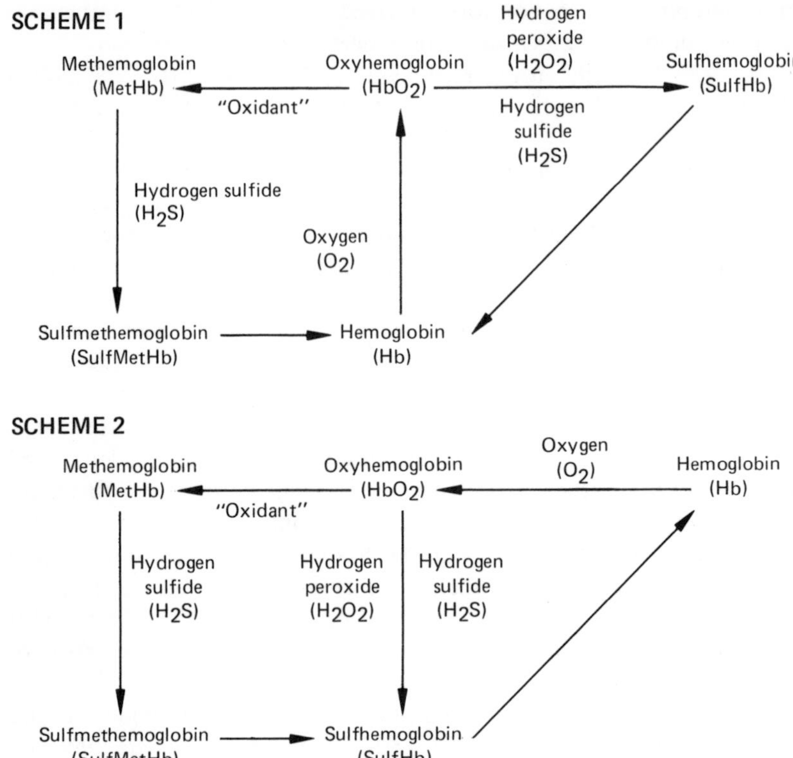

Figure 4-2. Hypothetical schemes for a suspected interrelationship between sulfhemoglobin and sulfmethemoglobin. Both pigments appear to be unstable and tend to regenerate hemoglobin.

Thus, there is a complex between the hydrosulfide anion and the ferric heme iron of methemoglobin that is distinctly different in its properties from "sulfhemoglobin." This complex, sulfmethemoglobin, forms almost instantaneously, but it decays with a half-life of more than 2 hr to oxyhemoglobin. Sulfhemoglobin also decays to oxyhemoglobin, suggesting an undefined relationship with sulfmethemoglobin. Recent[153] work with sulfmyoglobins and sulfhemoglobins indicates that there may be a wide variety of ferrous and ferric derivates and that their chemistry and interrelationships may be extremely complex. It is interesting that the ionic forms of cyanide, sulfide, and azide bind to methemoglobin whereas the respective undissociated acids are the more active inhibitors of cytochrome oxidase.

Animal experiments In 1956, Gunter[126] was apparently the first to explore systematically the use of methemoglobin-generating chemicals as antidotes to acute sulfide poisoning. Not only do cyanide and sulfide have similar toxicologic effects, but also they both inhibit cytochrome oxidase and form complexes with methemoglobin. Therefore, it was natural to try with sulfide the therapeutic approach that had been used so successfully with cyanide. Gunter[126] showed that prophylactic sodium nitrate prevented death in rabbits and dogs acutely poisoned with hydrogen sulfide. Sodium nitrite or methylene blue given during the "paralytic" stage to animals poisoned with six times the lethal dose of ammonium sulfide saved them from death. He also reported that pyrogallol prevented death from acute sulfide poisoning.

Methylene blue and pyrogallol are weak methemoglobin-generating agents at best;[290, 291] but sodium nitrite is a classical cyanide antagonist. Intravascular and intraerythrocytic methemoglobin can trap free cyanide in the biologically inactive form of cyanmethemoglobin. An analogous mechanism is almost certainly at work in the effects of methemoglobinemia on the course of acute sulfide poisoning.[288]

The findings of Gunter[126] were confirmed by Scheler and Kabisch[273] in mice exposed to lethal atmospheres of hydrogen sulfide. The mean survival time of a group of control mice exposed to 0.118% hydrogen sulfide by volume was 6.1 ± 2.2 min. Pretreatment with sodium nitrite 20 min before exposure increased the survival time of mice significantly and in proportion to the dose of nitrite. Similarly, the LC_{50} (lethal concentration

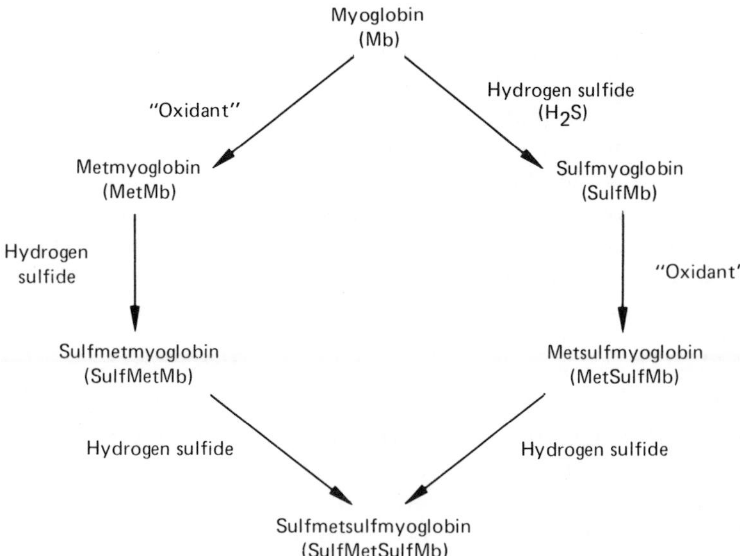

Figure 4-3. Hypothetical scheme for the interactions of sulfide with ferrous and ferric myoglobin. Sulfmyoglobin involves an attachment of sulfide to the heme group but not to the iron, whereas sulfmetmyoglobin involves binding of sulfide to the heme iron. According to Peisach,[249] both types of binding can occur simultaneously.

for 50% of a group of mice after 20-min exposures) was 0.043% by volume for control animals, but this increased to 0.138% in mice given 2.2 mg of sodium nitrite. Hydrogen sulfide poisoning was reversed in mice by injecting nitrite into their tail veins with cannulas when the animals were severely poisoned.[273]

Smith and Gosselin[287] demonstrated significant protective effects of methemoglobinemia against acute sulfide poisoning in armadillos, rabbits, and mice. The degree of protection was related to the circulating titer of methemoglobin irrespective of the agent that was used to generate it. Nitrite, hydroxylamine and *p*-aminopropiophenone can generate equivalent peak concentrations of methemoglobin *in vivo*, but at widely different doses and at different times after injection. The degree of protection afforded to mice was the same in each case when sulfide was given at the peak of the methemoglobinemias.[284]

When mice were continuously exposed to gas concentrations of 0.07%, 0.10%, or 0.19% by volume, the largest protection index (mean survival time of methemoglobinemic mice/mean survival time of control mice) was obtained with the intermediate concentration of hydrogen sulfide. Presumably, at high concentrations the fixed pool of methemoglobin is very rapidly saturated, whereas at low concentrations the methemoglobin pool becomes increasingly superfluous in the face of active endogenous detoxication.[287]

The ratio of the LD_{50} in methemoglobinemic mice to the LD_{50} of control mice was greater with sodium cyanide than with sodium sulfide when the chemicals were given intraperitoneally.[288] A marginal protective effect was also observed for sodium azide. These findings were in accord with the relative affinities of methemoglobin for the three anions, i.e., cyanide was bound the most tightly and azide the least. Nevertheless, more sulfide seemed to be inactivated *in vivo* by the standard dose of nitrite than cyanide. Some evidence suggested that nitrite-generated methemoglobin had more binding sites available for sulfide than for cyanide, or that it had sites exclusively available to sulfide in addition to the ferric heme iron.[284, 288] In extensive studies on the binding of sulfide to various methemoglobins *in vitro*, however, this observation could not be confirmed. Only one sulfide atom appeared to be bound per iron atom (Kruszyna, H., Kruszyna, R., and R. P. Smith, 1976, unpublished observations). Perhaps nitrite protects against sulfide poisoning by mechanisms in addition to methemoglobin formation.

As indicated above, the rate of reaction of sulfide with disulfide bonds under physiologic conditions is much more rapid than the rate of reaction of cyanide with such bonds. Accordingly, mice can be protected against acute sulfide poisoning by pretreating them with oxidized glutathione; no such protective effect was observed with cyanide or azide.[286] The effect of oxidized glutathione is additive to that of methemoglobinemia, and presumably it could be imparted by a wide variety of disulfide compounds. Cobaltous chloride also protects mice against acute poisoning by sulfide and by cyanide.[285] This effect is probably due to direct chemical inactivation of these toxic anions by cobalt.

The remaining widely recommended antidote to sulfide poisoning is oxygen given by positive pressure or as artificial respiration. This recommendation appears to be largely empirical. For example, Evans[103] recommended artificial respiration instead of the induction of methemoglobinemia but did not report confirmatory evidence of either point. As evaluated in mice, oxygen (100% at 1 atm) had no significant protective or therapeutic effect against death by intraperitoneal sodium sulfide when compared with animals maintained under air at 1 atm. Nor did oxygen modify in any way the marginal protective effect of thiosulfate or the much larger protective effect of nitrite. At the same time, an antidotal effect of nitrite against intraperitoneal sodium sulfide was also demonstrated in rats.[289]

The antagonistic effects of methemoglobinemia and the failure of oxygen as an antidote are in keeping with the suspected key biochemical lesion in acute sulfide poisoning, namely an inhibition of cytochrome oxidase. As pointed out above, this lesion represents an inability to utilize oxygen when its availability and transport are not impaired. The apparent efficacy of artificial respiration in sulfide poisoning was probably due to coincidental termination of the exposure with spontaneous recovery or to the resulting increase in pulmonary excretion of hydrogen sulfide. Oxygen should never be withheld, however, because of the possibility of ensuing pulmonary edema (see discussion below).

In summary, induced methemoglobinemia affords significant protective and antidotal effects in acute experimental sulfide poisoning. Oxidized glutathione, other simple disulfides, and cobaltous chloride protect animals against the acute lethal effects of sulfide, presumably by inactivating it chemically. Oxygen at atmospheric pressure appears to be devoid of significant antagonistic effects with respect to the acute syndrome. Because the endogenous detoxication of sulfide proceeds rapidly, antidotes are best reserved for the critically ill patient; however, the therapeutic induction of methemoglobinemia has been employed successfully in at least one severe human intoxication.[301]

CHRONIC AND SUBACUTE POISONING BY SULFIDES

Investigations into the chronic toxicity of sulfides are almost nonexistent. The results of one 90-day continuous inhalation exposure were published only as an uninformative abstract.[265] The mean lethal dose of sulfide varies greatly with the rate of administration. When it is given slowly by intravenous infusion, animals can tolerate several multiples of the mean lethal dose as determined by rapid injection of a single bolus. It is therefore likely that most animal species have a large endogenous capacity to detoxify sulfide. Indeed, Bittersohl[37] states that experiments to date have failed to confirm chronic or cumulative effects of hydrogen sulfide. Low cumulative toxicity can also be predicted by analogy with cyanide, which appears to be virtually devoid of chronic effects.[132] According to Fyn-Djui,[115] however, exposure of rats to 10 mg/m^3 for 12 hr/day over 3 months resulted in functional changes in the central nervous system and morphologic changes in the brain cortex. The evaluation of chronic or cumulative effects should be a high priority item in future research on this common chemical.

Lendle[188] gave sodium sulfide and sodium cyanide to guinea pigs by slow intravenous infusion. The concentrations of the solutions were adjusted so that fatal respiratory arrest always occurred between 20 and 100 min after the start of the infusion. From the amount infused and the acute lethal dose given as a single bolus he calculated that guinea pigs are capable of detoxifying 85% of the single lethal dose of sulfide and 90% of the single lethal dose of cyanide each hour. Presumably this effect represents metabolic detoxication, but there was no direct evidence to support this. When common cyanide antagonists were tested for their ability to increase significantly the lethal dose of sulfide as given over the same infusion time to both treated and control animals, none was effective. Neither protective nor antidotal effects were obtained in experiments with oxygen, methemoglobin-generating chemicals, iron citrate, zinc sodium edetate chelate, zinc lactate, or various cobalt chelates. These experiments do not imply that the character of the lethal mechanism of sulfide was different under these conditions. They only demonstrate that endogenous detoxication assumes greater importance when sulfide is given slowly.

Irritant Effects of Hydrogen Sulfide

Pulmonary edema The human experience with the long-term irritant effects of hydrogen sulfide probably constitutes a larger body of information than the experimental data, which are now somewhat antiquated.[268, 269] Table 4-1 summarizes the gas concentrations that produce "subacute" as contrasted with acute intoxications in six species. The "subacute" syndrome refers to the production of pulmonary edema and other signs of irritation as opposed to the fulminating asphyxia without pulmonary involvement in the acute syndrome. The data indicate that all six species are remarkably uniform in their susceptibility to hydrogen sulfide, i.e., they show very little in the way of species differences. This uniformity may even extend to insects. At least, the order of decreasing toxicity for hydrogen cyanide, hydrogen sulfide, chlorine, sulfur dioxide, and ammonia was the same in mice, rats, and houseflies.[332]

The signs exhibited in the "subacute" exposures of Table 4-1 included salivation, irritation of the eyes and respiratory tract, and dyspnea. Death intervened in about 10 to 18 hr when dogs were exposed continuously to 0.015% to 0.035% hydrogen sulfide by volume. At autopsy these animals differed considerably from those that died in the acute syndrome. Fluid, sometimes blood-tinged, was usually present in the pleural cavity. The lungs filled the cavity and were mottled in color from dark purple to white. They were also soggy and pitted on pressure. Creamy frothy material exuded from the bronchioles. Clearly, these animals expired from massive pulmonary edema. The mechanism of the edemagenic activity is unknown. Although it is usually ascribed to nonspecific irritation, the possibility that sulfide interferes with the metabolic function of the alveolar lining or the lung capillaries has not been explored.

Sulfide-induced tolerance to other edemagenic agents Interestingly, hydrogen sulfide is included among a number of sulfur compounds that protect mice against death from respiratory exposures to ozone or nitrogen dioxide. The simultaneous inhalation of 1-hexanethiol, methanethiol, dimethyl disulfide, hydrogen polysulfide, di-*tert*-butyl disulfide, benzenethiol, thiophene, and hydrogen sulfide protected mice against death by ozone or nitrogen dioxide. However, the protective effects of the last two sulfur compounds were significantly greater than those of any of the others. Protection against ozone and nitrogen dioxide was also imparted by the injection of thiourea derivatives several days before exposure.[105] Again the mechanism of this protective effect is unknown, but the observations suggest that some critical balance between free sulfhydryl groups and disulfide bonds may be important.

As stated above, there is a glaring gap in the present knowledge about the effects of chronic sulfide poisoning. Although there is some evi-

Table 4-1. Concentrations of hydrogen sulfide resulting in subacute or acute intoxication syndromes in various species[a]

Species	Subacute syndrome (% hydrogen sulfide by volume)	Acute syndrome (% hydrogen sulfide by volume)
Canaries	0.005-0.020	≥0.02
White rats	0.005-0.055	≥0.05
Dogs	0.005-0.065	≥0.06
Guinea pigs	0.010-0.075	≥0.075
Goats	0.010-0.090	≥0.090
Humans	0.010-0.060	0.06-0.1[b]

[a] Derived from Sayers *et al.*, 1923.[268] See Chapter 5 for definitions of acute and subacute syndromes.

[b] Immediately lethal.

dence to predict a low order of cumulative toxicity, the possibility that sulfhemoglobinemia might be a prominent sign after long exposures should be examined. The key lesion on subacute exposure of animals and humans to hydrogen sulfide gas is pulmonary edema. Paradoxically, the simultaneous inhalation of hydrogen sulfide protects animals against the lung edema caused by "oxidant" pollutants such as ozone and nitrogen dioxide.

TOXICOLOGY OF SOME SIMPLE MERCAPTANS

The acute intoxication syndrome produced by sulfide appears to be unique among simple sulfur-containing compounds. Although some simple mercaptans are appreciably toxic, they appear to mimic only the subacute syndrome induced by hydrogen sulfide. As shown in Table 4-2, hydrogen sulfide was five times more acutely toxic to rats than dimethyl disulfide and 10 times more toxic than methyl mercaptan. Although all of the aliphatic sulfides in Table 4-2 appeared to produce death in pulmonary edema, systemic effects cannot be ruled out.[199] Dimethyl disulfide has recently been identified as an attractant pheromone in hamster vaginal secretions.[281]

Coma occurred in 50% of a rat population within 15 min of exposure to 0.16% methanethiol, 3.3% ethanethiol, or 9.6% dimethyl sulfide by volume. All animals recovered consciousness within 30 min. These mercaptans potentiate the coma produced by ammonium salts in rats. Some of them have been detected in the urine and breath of patients with massive hepatic necrosis.[341] Methanethiol and other alkylthiols appear to be at least partly responsible for fetor hepaticus, the unpleasant breath odor of comatose patients with severe liver disease. Methanethiol is a potent inhibitor of rat liver mitochondrial respiration and may react with cytochrome oxidase.[328] As noted above, ethyl mercaptan, and presumably other simple mercaptans, do not generate sulfhemoglobin under conditions in which hydrogen sulfide does.

Table 4-2. Concentrations of hydrogen sulfide and simple mercaptans lethal to rats after 10- to 20-min exposures[a]

Compound	Percent by volume
Hydrogen sulfide	0.1
Dimethyl disulfide	0.5
Methyl mercaptan	1.0
Dimethyl sulfide	5.4

[a] Derived from Ljunggren and Norberg, 1943.[199]

It has been suggested that the acute toxicity of phenylthiourea is due to the release of hydrogen sulfide *in vivo*, but that assertion seems most unlikely.[123] One should anticipate sulfide poisoning from exposure to any soluble sulfide salt. Barium sulfide is an exception because symptoms of barium poisoning predominate over those of sulfide.[123]

Similarly, the toxic effects of carbon disulfide have been ascribed in part to its metabolism to hydrogen sulfide. Except for the circumstance that workers in the viscose rayon and other artificial fiber industries are often concurrently exposed to both chemicals, there is no convincing evidence to support this. The intoxication syndromes induced by the two chemicals are quite dissimilar. Instead of hydrogen sulfide, carbon disulfide is metabolized to carbonyl sulfide and an unknown reactive form of sulfur.[72, 123]

5

Effects on Humans

There is probably no odor more readily identifiable to the average individual than that of hydrogen sulfide. Very low concentrations of the gas, which may be encountered almost anywhere, are easily detected by olfaction. Hydrogen sulfide may evolve naturally wherever organic matter undergoes putrefaction and it is released into the air as a worthless byproduct of countless industrial processes. To most people, the odor of hydrogen sulfide is nothing more than an unpleasant nuisance. Generations of chemistry students have learned to accept its lingering presence in the laboratory air with little concern. Yet, at higher concentrations, hydrogen sulfide is a deadly poison. It is nearly as toxic as hydrogen cyanide, and its action may be as rapid. Deaths due to hydrogen sulfide intoxication, which are reported regularly, are usually associated with exposure under occupation-related circumstances. Occasionally, however, they result from contact with the naturally occurring gas or, rarely, are consequent to accidental release of industrially generated hydrogen sulfide into the community.

The characteristic olfactory response to hydrogen sulfide is an important aspect of its toxicology. Its typical "rotten-egg" odor is detectable by olfaction at very low concentrations [0.035 μg/liter (0.025 ppm)] in the air. Exposures to these low concentrations have little or no importance to human health. Thus, this olfactory response is a safe and useful warning signal that a hydrogen sulfide source is nearby. However, at higher concentrations [$\geq$280 μg/liter (200 ppm)], hydrogen sulfide is distinctly dangerous. It exerts a paralyzing effect on the olfactory apparatus, effectively neutralizing the olfactory warning signal. Table 5-1 lists some of the most significant effects of hydrogen sulfide exposure on humans, together with estimates of the concentrations at which these effects may be expected.

Hydrogen sulfide is an irritant gas. Its direct action on tissues induces local inflammation of the moist membranes of the eye and respiratory tract.[128] The dry surfaces of the skin are seldom affected by gaseous hydrogen sulfide, nor does the gas penetrate the intact skin to any great extent.[338] However, Aves[15] noted from his experience that hydrogen sulfide noticeably retards the healing of minor skin wounds. When inhaled, hydrogen sulfide exerts its irritative action more or less uniformly throughout the respiratory tract, although the deeper pulmonary structures suffer the greatest damage. Inflammation of these

Table 5-1. Effects of hydrogen sulfide exposure at various concentrations in air

Effect	Concentration	
	μg/liter	ppm
Approximate threshold for odor[338]	0.14-0.28	0.1-0.2
Offensive odor[338]	4.2-7	3-5
Threshold limit value[a,12]	14	10
Threshold of serious eye injury (gas eye)[b,8]	70-140	50-100
Olfactory paralysis[8,269]	210-350	150-250
Pulmonary edema, imminent threat to life[128]	420-700	300-500
Strong nervous system stimulation, apnea[128]	700-1,400	500-1,000
Immediate collapse with respiratory paralysis[128]	1,400-2,800	1,000-2,000

[a] In the United States, permissible concentrations of hydrogen sulfide in the workroom air have been promulgated by the U.S. Department of Labor, Occupational Safety and Health Administration (OSHA), and published in the *Federal Register* (1910.93).[322] These include an "acceptable ceiling concentration" of 28 μg/liter (20 ppm) and an "acceptable maximum peak above the acceptable ceiling concentration for an 8-hr shift" of 70 μg/liter (50 ppm) for "10 minutes once only if no other measurable exposure occurs." This is interpreted by OSHA to mean that an employee's exposure to hydrogen sulfide may not exceed, at any time during an 8-hr shift, 28 μg/liter (20 ppm) of hydrogen sulfide, except once, for a 10-min period, during which interval the concentration may be as high as 70 μg/liter (50 ppm). The OSHA regulations do not list an 8-hr weighted average (TWA) for hydrogen sulfide; however, a TWA of 14 μg/liter (10 ppm) has been recommended by the American Conference of Governmental Industrial Hygienists (ACGIH).[12]

[b] See "Effects of Hydrogen Sulfide on the Eyes" beginning on p. 61 of this chapter for definition.

structures may appear as pulmonary edema. Also, hydrogen sulfide, which is readily absorbed through the lung, can produce fatal systemic intoxication if inhaled at sufficiently high concentrations. Most authors have found it convenient to categorize hydrogen sulfide poisoning under three rubrics according to the nature of the dominant clinical signs and symptoms. These rubrics are acute, subacute, and chronic poisoning.

The term "acute hydrogen sulfide intoxication" has been used most often to describe episodes of systemic poisoning characterized by rapid onset and predominance of signs and symptoms of nervous system involvement. Hydrogen sulfide concentrations in the range of 700 to 980 μg/liter (500 to 1,000 ppm) will usually produce manifestations of acute intoxication after a few minutes of exposure.[269] At higher concentrations, effects appear even more rapidly.

In "subacute intoxication", signs and symptoms of eye and respiratory tract irritation prevail. Although subacute intoxication can follow a single, brief but intense exposure (such as a jet of gas into the eye), subacute effects are more often related to exposures of several hours. At hydrogen sulfide concentrations of about 140 to 210 μg/liter (50 to 100 ppm), eye irritation becomes noticeable in several minutes, and respiratory tract irritation without coughing may occur in approximately 30 min.[269] Pulmonary edema, the major threat to life associated with the local irritant properties of hydrogen sulfide, may be precipitated by a few hours of exposure to concentrations of several hundred micrograms per liter or may be a sequela of acute, nonfatal intoxication.[280]

"Chronic hydrogen sulfide intoxication" is applied by some investigators to a prolonged state of symptoms resulting from single or repeated exposures to concentrations of hydrogen sulfide that fail to produce clear-cut manifestations of either acute or subacute illness. Not all authors accept chronic hydrogen sulfide poisoning as a condition that is distinct from mild acute and/or subacute poisoning.[8, 15]

Finally, it must be emphasized that so-called acute intoxication is frequently followed by evidence of local irritation, both of the eyes and the lung. Occasionally, this local irritation leads to the development of pulmonary edema (see "Effects on the Respiratory Tract" beginning on p. 64 this chapter).

The Subcommittee on Hydrogen Sulfide regards the terms acute, subacute, and chronic as both imprecise and misleading when used to describe the effects of hydrogen sulfide exposure. Rather than to abandon these frequently used terms altogether, the subcommittee offers the following clarifying definitions:

Acute intoxication: Effects of a *single* exposure to *massive* concentrations of hydrogen sulfide that rapidly produce signs of respiratory distress. Concentrations approximating 1,400 μg/liter (1,000 ppm) are usually required to cause acute intoxication.

Subacute intoxication: Effects of *continuous* exposure to *mid-level* [140 to 1,400 μg/liter (100 to 1,000 ppm)] concentrations of hydrogen sulfide. Eye irritation (gas eye, see "Effects of Hydrogen Sulfide on the Eyes" beginning on p. 61) is the most commonly reported effect, but pulmonary edema (in the absence of acute intoxication) has also been noted.

Chronic intoxication: Effects of *intermittent* exposures to *low to intermediate* concentrations [70 to 140 μg/liter (50 to 100 ppm)] of hydrogen sulfide, characterized by "lingering," largely subjective manifestations of illness.

ACUTE POISONING

In its acute form, hydrogen sulfide poisoning is a systematic intoxication—the result of the gas's action on the nervous system. The toxic effects of hydrogen sulfide are believed to be a consequence of reversible inactivation of cellular cytochrome oxidase that results in inhibition of aerobic metabolism.[235] Only in the form of free, unoxidized gas in the bloodstream can hydrogen sulfide exert its systemic effects. Because it is very rapidly oxidized in the blood to harmless and easily eliminated sulfates, hydrogen sulfide may be considered a noncumulative poison.

The quantitative relationships between hydrogen sulfide concentrations in the air and the nature and severity of systemic effects were first reported in 1892 by Lehmann.[186] He exposed men to concentrations of hydrogen sulfide ranging from 140 to 700 μg/liter (100 to 500 ppm), which produced severe poisoning. The effects were similar to those observed in animals at similar levels of exposure. Some years later, in 1925, Sayers *et al.*[269] exposed "some men" for short periods to low levels of hydrogen sulfide in the air. They concluded that men could not be used safely as experimental subjects because of the possible injury to the lungs and the narrow margin between consciousness and unconsciousness. Their results, together with those of Lehmann, produced enough data to predict the reaction of men to higher concentrations.

Also in 1925, Haggard[128] reported his experiments with hydrogen sulfide using dogs as experimental subjects. He emphasized that his results were in complete agreement with the observations reported by Lehmann. He correctly identified the role of hydrogen sulfide as a highly dangerous asphyxiant, clearly described its effects on the nervous control of respiration, and explained the underlying mechanisms. His notions on the subject remain essentially valid today and have been drawn upon heavily in the following paragraphs.

At concentrations in air exceeding ~700 μg/liter (500 ppm), hydrogen sulfide must be considered a serious and imminent threat to life.[128, 269] Between 700 and 1,400 μg/liter (500 and 1,000 ppm), hydrogen sulfide has permanent effects on the nervous system. Although a number of organs and tissues, including the heart and skeletal muscles, respond, this action is most significantly expressed through its effect on the chemoreceptors of the carotid body.[136] This stimulation results in excessively rapid breathing (hyperpnea), which quickly leads to depletion of the carbon dioxide content of the blood (acapnia). This, in turn, gives way to a period of respiratory inactivity (apnea). If depletion has not progressed too far, carbon dioxide may reaccumulate to the point where spontaneous respiration is reestablished. If spontaneous recovery does not occur and artificial respiration is not immediately provided, death from asphyxia is the inevitable conclusion. At about 2,100 μg/liter (1,500 ppm), the course of events is the same except that the reaction is more intense. At concentrations of $\geq$2,800 μg/liter (2,000 ppm), breathing becomes paralyzed after a breath or two; in Haggard's words, "the victim falls to the ground as though struck down." Generalized convulsions frequently begin at this point.

This form of respiratory failure is not related to the carbon dioxide content of blood. Rather, hydrogen sulfide exerts a direct paralyzing effect on the respiratory center.[133] According to Haggard, breathing is never reestablished spontaneously following this hydrogen sulfide-induced paralysis of respiration. However, because the heart continues to beat for several minutes after respiration has ceased, death from asphyxia can be prevented if artificial respiration is begun immediately and is continued until the hydrogen sulfide concentration in the blood decreases as a result of pulmonary excretion of the gas. Once sufficient hydrogen sulfide is excreted—after several minutes—normal respiration is usually reestablished.

Victims of acute hydrogen sulfide poisoning who recover usually do so promptly and completely. Sequelae occasionally follow acute poisoning, but the frequency with which they appear is difficult to estimate accurately from published accounts. It is usually only the "interesting case" that finds its way into the medical literature. What can be said is that the sequelae vary widely in nature and severity. Some examples are given later in this chapter.

Tolerance to the acute effects of hydrogen sulfide is not acquired, as it is with sulfur dioxide.[8] Instead, authors often mention the term "hypersusceptibility" in connection with the response of some individuals to repeated exposures to the gas.[8, 183, 324] Unfortunately, the meaning of that term, as used among publications, does not consistently conform to the accepted definition, which according to *Dorland's Illustrated Medical Dictionary* (Ref. 87, p. 744), is "a condition of abnormally increased susceptibility to poisons . . . which in the normal individual are entirely innocuous." To Ahlborg,[8] hypersusceptibility refers to an acquired, intense aversion to the odor of hydrogen sulfide. Others suggest a lessening of resistance to the toxic action of the gas.

In view of hydrogen sulfide's frequent appearance as a consequence of man's activities and nature's whims, together with its especially malevolent toxic nature, it is not surprising that numerous accounts of acute hydrogen sulfide poisoning have accumulated over the years. Mitchell and Davenport,[220] in their noteworthy paper of 1924, reviewed much of the very early published work on hydrogen sulfide poisoning (see Appendix II).

Occupational Exposures

In the United States between 1925 and 1930, several published reports called attention to the role of hydrogen sulfide as an occupational hazard of major importance.[15, 221, 338] The authors of each report cited the introduction into the United States of high-sulfur Mexican crude oil as the cause of numerous cases of hydrogen sulfide poisoning among oil field workers. Mitchell and Yant, in their report published in 1925 by the U.S. Department of the Interior's Bureau of Mines,[221] reviewed insurance company and plant records for accidents attributable to hydrogen sulfide exposure during the refining of high-sulfur crudes. They cited 58 cases of asphyxia and 99 cases of irritation that had taken place in the "few years" since Mexican crude was introduced. Some fatalities occurred, but the exact number was not provided by the authors.

In 1929, Aves[15] published a vivid account of hydrogen sulfide poisoning in the Texas oil fields. He noted that although hydrogen sulfide "accidents had occurred spasmodically" it was not until high-sulfur Mexican oils were introduced that the problem became severe. All natural life, such as birds and rabbits, disappeared from the oil fields. He estimated that during a 2-year period, 15 to 30 deaths occurred as a result of hydrogen sulfide poisoning. His final paragraph is especially descriptive of the situation as he perceived it:

> It is quite a surprise to one to find that the old "rotten egg" gas of our laboratory days is as toxic as hydrocyanic acid, and that it is coming from nature's laboratory three thousand feet [915 meters] underground in such concentrations, and millions of cubic feet of it every day. All the lead paint in the area has turned black. The nickel on our automobiles and all silver money does likewise.

In 1930, Yant[338] published a detailed account of the importance of hydrogen sulfide as an industrial hazard. His assessment of the then current awareness of the problem is informative:

> The lack of general familiarity with hydrogen sulphide poisoning among industrial surgeons . . . has been due to the fact that until 10 or 15 years ago cases of poisoning were rather unusual and did not constitute a major industrial health hazard. Since then, hydrogen sulphide poisoning has become a major hazard in certain industries. . . .

Yant then proceeded to enumerate the principal industrial sources of hydrogen sulfide. He believed that the most important source was the petroleum industry, as a direct result of the importation of the high-sulfur Mexican crude oil.

In 1951, Ahlborg[8] published a comprehensive article on hydrogen sulfide poisoning in Sweden's shale oil industry. He discussed the relative health hazards of the various chemicals encountered in the industry and concluded that hydrogen sulfide was by far the most toxic. Since the shale being worked contained approximately 7% sulfur, considerable hydrogen sulfide was produced in the distillation process. The author stated that beginning in 1945, hydrogen sulfide levels were determined in 107 work areas throughout the plant. Nine locations had more than 840 μg/liter (600 ppm) hydrogen sulfide; an additional 11 locations had more than 210 μg/liter (150 ppm); and the majority of the remaining areas had less than 28 μg/liter (20 ppm). The highest concentrations were found in areas where men worked only occasionally.

Between 1943 and 1946, 58 workers suffered acute hydrogen sulfide poisoning with unconsciousness; 14 were hospitalized but there were no deaths. There were also several cases of acute poisoning without unconsciousness. Among all these cases, 15 suffered sequelae of one sort or another. Of these, 12 exhibited the following characteristics: a history of repeated poisoning; symptoms of sequelae developing immediately or soon after acute poisoning; and a period of recovery of about 1.5 months. Sequelae consisted of both neurasthenic symptoms, such as headache, fatigue, dizziness, irritability, anxiety, and poor memory, and otoneurologic manifestations, including disturbances of equilibrium, nystagmus, and deviation from normal gait and arm movements. Most of these patients also complained of considerable increase in sensitivity (aversion) to the odor of gas of any type—even to pure gasoline vapor. In none of the seven cases with sequelae described in detail by Ahlborg did the patient appear to have suffered a severe episode of acute intoxication. In at least one case, the patient never lost consciousness. From Ahlborg's series of cases, it appears that sequelae can follow acute hydrogen sulfide intoxication without an intervening episode of hypoxemia.

Other reports of acute hydrogen sulfide poisoning have also stressed the occurrence of sequelae. In 1955, Kaipainen[160] described the case of a 48-year-old farmer who collapsed from hydrogen sulfide intoxication while shoveling manure. After resuscitation 2 hr later, he continued to have convulsive seizures. His electrocardiogram (ECG) exhibited changes suggestive of myocardial infarction. Kidney lesions were indicated by the elevation on nonprotein nitrogen levels in the blood and by the presence of red blood cells and hyaline casts in the urine. The patient recovered after 1.5 months, but he retained a slight, persistent dizziness.

Hurwitz and Taylor[144] reported the case of a 46-year-old sewer worker who, upon descending into a manhole to investigate "the foulest smell I have ever come across," began to feel faint. He tried to ascend but fell back unconscious. Thirty minutes elapsed before he could be brought out into fresh air . By this time, he was cyanotic, his teeth were clenched, and he was having generalized "toxic spasms," each lasting 10 to 20 sec and occurring every 2 min. Artificial respiration was necessary after each convulsion. The patient recovered very slowly. A week after the accident he could move and speak only with great effort. A month afterward he still had neurologic disabilities. His ECG showed evidence of a small anterolateral infarct and a right bundle-branch block. Three months after the accident he was walking normally but suffered anginal pain upon exertion.

A third case involving sequelae was reported in 1966 by Kemper.[168] A 31-year-old refinery worker was discovered unconscious, apneic, and deeply cyanotic near a spill of diethanolamine charged with hydrogen sulfide. He had apparently been unconscious for several minutes before his supervisor, working from the upwind side, was able to drag him clear of the area and begin artificial respiration and oxygen administration. The patient was hospitalized and ultimately survived a stormy course of acute illness, which included intermittent respiratory failure, cardiovascular collapse, violent convulsions, bilateral bronchopneumonia, and electrocardiographic evidence of myocardial ischemia. Fifteen days after admission, he was discharged with no abnormal physical signs. His ECG gradually returned to normal. For several months, he suffered mild depression and lassitude. Amnesia for the day of the accident remained complete a year later.

In all three of these cases, prolonged hypoxia of vital tissues might have explained the pronounced sequelae that were observed. Certainly, the condition common to each event was the considerable period of unconsciousness. Numerous other reports of sequelae following acute hydrogen sulfide intoxication can be found in the medical literature. Most resemble the three cases just cited, in which serious poisoning with unconsciousness preceded the appearance of sequelae. The observations reported by Ahlborg[8] depart somewhat from this pattern for reasons that are not clear.

Several reports have focused upon autopsy findings in victims of fatal occupational hydrogen sulfide poisoning. In 1961, Breysse[43] wrote of a workman in a poultry feather-fertilizer plant who went into a nearby marsh to repair a leak in a cooker-outfall line. The effluent from the feather cooker had been casting an obnoxious odor throughout the vicinity, prompting numerous complaints. About 15 min after the man had left, his worried coworkers found him dead, slumped over the outfall line. The report of the autopsy emphasized the dusky, cyanotic color of the victim's skin and brain, as well as marked bilateral hemorrhagic pulmonary edema and cerebral edema. An examination of the deceased's blood confirmed the presence of hydrogen sulfide. A week after the accident, an air-sampling survey revealed that the hydrogen sulfide near the leak in the outfall line reached concentrations as high as 5,600 μg/liter (4,000 ppm) during the normal cooking process.

Three years later, another very similar case was reported by Larson *et al.*[183] The victim was a 34-year-old employee of another poultry feather-fertilizer plant. Again, foul odors emanated from an outfall line that dipped into a nearby bay. The victim investigated the problem and was later found dead near the outfall pipe. The autopsy revealed findings that were similar to those of Breysse, except for the absence of cerebral edema. What especially intrigued the authors was a "light purple" color of the brain, which faded rapidly following immersion in formalin. They speculated that the sulfhemoglobin (see Chapter 4) remaining in the brain after inadequate embalming was the cause of the purple color. A test of the blood for sulfur was inconclusive, although the test for hydrogen sulfide was positive. The authors emphasized that the victim had experienced repeated exposures to hydrogen sulfide. They theorized that he had become accustomed to the local irritative properties and therefore was numb to the odor of the gas. This theory is not supported by the balance of the literature on acute hydrogen sulfide poisoning.

A third autopsy case was reported by Adelson and Sunshine[7] in 1966. A 35-year-old workman descended into a 4.6 m-deep sewer to collect water samples. When he failed to answer their call, two of his coworkers, ages 21 and 25, went in after him, while a third man held a rope with which to pull the victim to the surface. The two would-be rescuers themselves succumbed. It was not until firemen with gas masks arrived that all three were finally brought back to the surface.

Resuscitation of the victims failed, and all three were pronounced dead on arrival at the hospital. The autopsy reports noted that the victims reeked of a "rotten egg" odor and their skin exhibited a grayish green cyanosis. Internally, each had hemorrhagic pulmonary edema and an unusual greenish cast to his blood and brain. The blood contained high concentrations of hydrogen sulfide. The authors speculated that the origin of the greenish color was not sulfhemoglobin, since significant quantities of sulfhemoglobin had never been found in the blood of hydrogen sulfide-induced fatalities before the onset of decomposition. They surmised that some other unstable hydrogen sulfide/hemoglobin complex was responsible for the unusual color. The authors discussed in some detail the environmental background of the tragedy. They noted that witnesses later recalled a "terrific odor" emanating from the manhole, whereas normally such odors were not associated with that area. An investigation revealed that a nearby factory, which converted fish oil and crude petroleum oil into gear lubricant, discharged about 225 kg of hydrogen sulfide into the sewer every 36 hr. Normally, the gas was dissolved in wash water at a low concentration, but the day preceding the accident, the neighborhood (and factory) water supply had been shut off for 2.5 hr to allow for repair of a fire hydrant. The authors felt that this shutoff had upset the usual gas to water ratio, and, aided perhaps by acid discharge from other factories, had led to the liberation of hydrogen sulfide at high concentrations. The authors pointed out that because hydrogen sulfide is heavier than air, it tends to accumulate in sewers, mines, and other enclosed subsurface atmospheres.

In 1968, in a letter to *The Lancet*, Winek *et al.*[336] briefly reported the autopsy of another hydrogen sulfide victim, a 55-year-old man who died in a coal-tar resin tank. The findings were edema of lungs and brain, along with chronic passive congestion of lungs, liver, and spleen. The authors pointed out that the man's physical examination a year earlier had shown no abnormalities. Air samples from the tank revealed hydrogen sulfide levels of 2,700 to 8,500 μg/liter (1,900 to 6,100 ppm). The presence of these lethal concentrations was not explained.

An unusual source of exposure to hydrogen sulfide, which nearly resulted in two fatalities, was reported by Milby[218] in 1962. The problem involved a No. 3 gas cylinder containing 4 kg of hydrogen sulfide under pressure of 1.8×10^5 kg/m^2. The valve on the cylinder had corroded to the extent that it could not be safely forced open under a laboratory hood. Three men were given the task of disposing of the cylinder. They chose a method commonly in practice at the time. This involved removing the cylinder to an isolated area, then puncturing its wall from a distance with a well-placed bullet from a high powered rifle, thus permitting the gas to escape harmlessly into the air. Accordingly, the cylinder was partially buried, and the three men moved upwind approximately 46 m. They then prepared to penetrate the cylinder with a shot from a .375-caliber magnum rifle. One man fired the rifle while the other two stood behind him. Almost instantaneously with the shot a white cloud of hydrogen sulfide raced toward the group. All turned to flee, but before they could escape, the cloud was upon them. The marksman and one other fell unconscious and ceased breathing. The third man attempted to aid his companions but, feeling faint himself, could be of little help. Members of a line crew who had been observing from a nearby hilltop rushed to their aid. One rescuer began mouth-to-mouth resuscitation, but quickly fell unconscious himself. Shortly after, he recovered spontaneously. The other rescuers then employed the arm-lift, back-pressure method to resuscitate the victims. The two men resumed breathing, but began to convulse violently. They did not regain consciousness until after their arrival at the hospital approximately 30 min later. Both men were treated with oxygen and recovered completely within a few days, although during the recovery period one man complained of headache and chest tightness and the other of slight numbness of the extremities. The author emphasized that neither man could recall having noted the odor of hydrogen sulfide at the time of the incident. This is consistent with the estimated concentration of hydrogen sulfide to which the men were exposed—about 2,800 μg/liter (2,000 ppm)—which would have caused instant olfactory nerve paralysis. Milby also pointed out that the would-be rescuer who attempted mouth-to-mouth resuscitation but collapsed should have first dragged the victim out of the contaminated area before attempting first aid. He stressed that this should be a cardinal rule in the rescue of unconscious hydrogen sulfide victims. The rescuer could also have been poisoned by the hydrogen sulfide exhaled from the victim during mouth-to-mouth resuscitation.

In 1964, Kleinfield *et al.*[170] reported another unusual source of hydrogen sulfide exposure. A plant that produced benzyl polysulfide regularly used sodium sulfhydrate in its production processes. One day, a pipe used to transfer sodium sulfhydrate ruptured, spilling the liquid sulfhydrate over the ground and into a nearby sewer. There it reacted with the acid sewage, releasing hydrogen sulfide from several sewer openings in the immediate vicinity. Several tons of caustic were quickly dumped into the sewer to halt hydrogen sulfide generation. Meanwhile, rescue operations were started to save the 12 plant employees who had become severely intoxicated. The 12 victims were all located within a 15-m radius of the main hydrogen sulfide source at the time of the accident. Forty other employees out of the 89 present also became somewhat ill but did not lose consciousness. Of the 12 severely affected victims, two died. All of the survivors recovered without sequelae, probably because the duration of exposure was brief. Only 2 of the severely affected survivors could recall having smelled the gas, but a majority of those mildly affected remembered the odor of hydrogen sulfide at the time of the accident.

Kleinfeld *et al.*[170] also provided details on several other noteworthy cases. One 42-year-old man was found sitting 8 m from a leaking sodium sulfhydrate tank. He was ashen gray, but still conscious and breathing. The rescue team administered oxygen by mask, but when they saw other men unconscious nearby, they took the mask away to attend to them. Upon returning 15 min later they found their original patient apneic. They failed to revive him even after moving him into fresh air and giving him 30 min of vigorous resuscitation efforts. This again stresses the importance of immediately moving victims from contaminated areas.

Another victim, a 60-year-old man who was within a 15-m radius of the tank during the accident developed severe respiratory distress but did not lose consciousness. In the hospital, it was discovered that he had pulmonary edema. He recovered completely within 3 days.

In their summary, Kleinfeld *et al.*[170] conceded that this accident and the chain of events that followed were unusual. They noted, however, that similar accidents can occur wherever sulfur-containing materials are stored or processed, and stressed that precautions to guard against the occurrence of such unfortunate events must be carefully planned. In conclusion, the authors stressed their belief that the staff of this plant had been inadequately prepared to cope with emergencies.

One plant that did meticulously guard against hydrogen sulfide emergencies was, according to Poda,[253] a heavy-water production plant near Terre Haute, Indiana. This plant and affiliated plants used a hydrogen sulfide dual temperature exchange process that created a potentially major source of exposure to hydrogen sulfide. With considerable foresight, the plant management instituted a comprehensive safety program at the beginning of the plant's operation in 1957. This program included extensive personnel training for emergency situations. All employees were trained to administer artificial respiration and other first aid measures, to select safe evacuation routes determined primarily by wind direction, and to use lead acetate paper for hydrogen sulfide detection. Outdoor workers were required to carry air canisters at all times, while indoor workers were provided access to compressed air in permanent outlets. In addition, workers going into any potential gas area were required to do so under the "buddy system." Poda called attention to some interesting experiences with hydrogen sulfide at this plant. One peculiar problem was that of workers coming from outside areas of gas leakage into the control room and then suddenly collapsing. An investigation revealed that the men's clothes contained pockets of trapped hydrogen sulfide that quickly expanded in the warmth of the control room. This resulted in sudden exposure to the gas. The problem was solved by placing a fan in the doorway of the control room in order to dissipate the entrapped gas. Another major source of concern involved the open ditches that carried tower effluent to nearby seepage basins. The heat of the afternoon sun tended to liberate hydrogen sulfide dissolved in the effluent, thus posing a danger to workers. Enclosure of the drainage system solved that problem.

Despite all precautions over the years, 42 employees at this plant and its affiliates were rendered unconscious from hydrogen sulfide inhalation, although none died. Most victims stated that they had not smelled the characteristic hydrogen sulfide odor before losing consciousness; rather, they had very briefly smelled a sickening, sweet odor. Workers who inhaled sufficient gas to cause staggering or loss of consciousness often developed a syndrome characterized by nervousness, dry cough, nausea, headache, and insomnia. Those who had inhaled gas in an amount that was insufficient to cause staggering were usually given "carbogen"† to breathe for 10 min. Poda reported that the only serious case in 15 years was that of a mechanic found unconscious and cyanotic without pulse or respiration and with total incontinence of bladder and bowels. He was revived by artificial respiration and treated for pulmonary edema and shock. He recovered after 3 days and suffered no sequelae.

The conclusions that Poda drew from his extensive experience with hydrogen sulfide are worthy of note. In his opinion, hydrogen sulfide can be handled safely in large quantities as long as sufficient forethought is given to operational design and personnel training. He believed the "buddy system" to be absolutely essential in preventing fatalities, because intermediate treatment of victims is often crucial to survival. Poda emphasized that the toxicity of hydrogen sulfide is real and insidious. He suggested a "rule of thumb": if a person can smell the "rotten egg" odor of hydrogen sulfide, he can escape from a gas-contaminated area. He warned against the "familiarity breeds contempt" attitude that so often leads to accidents. He pointed out that many of the hydrogen sulfide victims that he had observed had either neglected to use their masks when they smelled the gas or had forgotten to check their

---

† Although not stated by the author, the word "carbogen" usually refers to a gas mixture containing 95% oxygen and 5% carbon dioxide.

canisters periodically for air pressure, valve function, etc., resulting in malfunction in time of need. Poda disagreed with the investigators who reported that hypersusceptibility to hydrogen sulfide was a potential sequel to repeated exposures. He claimed that among the cases he observed, no such phenomenon had occurred.

Community Exposures

Acute hydrogen sulfide poisoning is not solely an occupation-related problem. Occasionally, accidents involving community exposures have also been reported. The most dramatic and serious of such events occurred in 1950 at Poza Rica, Mexico, a city of 22,000 citizens located about 210 km northeast of Mexico City. Poza Rica was then the center of Mexico's leading oil-producing district and the site of several field installations, including a sulfur-recovery plant. At 2 a.m. on November 24, 1950, this plant, newly outfitted with units for burning hydrogen sulfide-rich gas, was authorized to increase the flow of gas to the flare at the full design capacity. Shortly thereafter, a malfunction of the flare apparatus permitted large quantities of unburned hydrogen sulfide to be released into the atmosphere. When the plant shut down some 3 hr later, 320 persons had been hospitalized due to acute illness; 22 of them died.

At the request of the Mexican government, McCabe and Clayton[210] investigated the disaster thoroughly. They found that a pronounced low level temperature inversion had occurred in the Poza Rica area on the morning of the tragedy, a condition common to most air pollution disasters. A slight breeze had apparently directed the unit's effluent gases, which included substantial quantities of unburned hydrogen sulfide, toward a nearby area of flimsily built bamboo dwellings. Although these structures were well suited for providing ventilation in a sultry climate, they offered little resistance to the entrance of poisonous gases. Between 4:50 a.m. and plant shutdown at 5:10 a.m., the residents of the area were abruptly awakened by the gas, and a bedlam of confusion ensued, with many people collapsing while others tried to assist them.

Of the 320 persons hospitalized, the authors made a detailed study of 47 who were "under particularly close observation," presumably because of the severity of their illness. The most frequent symptom was loss of the sense of smell, which was experienced by all but one of the 47. More than half of these patients had lost consciousness. Many suffered signs and symptoms of respiratory tract and eye irritation, and nine exhibited manifestations of pulmonary edema. The authors commented on the absence of respiratory and digestive sequelae but noted that the effects of the nervous system were prolonged. Four of the 320 victims were considered by the authors to have developed neurologic sequelae. Two patients experienced neuritis of the acoustic nerve; one developed dysarthria; and the last patient exhibited "marked aggravation" of his pre-existing epilepsy. The duration of these sequelae was not reported.

There have been reports of other episodes of general atmospheric pollution by industrially evolved hydrogen sulfide, but, fortunately, none have approximated the severity of the Poza Rica incident. In Terre Haute, Indiana, in 1964, biodegradation of industrial wastes in a 14.5-ha lagoon caused the atmospheric concentration of hydrogen sulfide to reach 0.4 μg/liter (0.3 ppm). Resulting complaints from the public numbered 81, 41 of which were health related. The most frequently reported symptoms were nausea, loss of sleep, abrupt awakening, breathlessness, and headache.[145] In 1973, residents of Alton, Illinois experienced a hydrogen sulfide episode that resulted in 266 health-related complaints being filed with the Illinois Environmental Protection Agency. Most complaints were of breathlessness and nausea.[145] In a report prepared for the U.S. Department of Health, Education, and Welfare in 1969,[219] reference was made to the hydrogen sulfide pollution problem around kraft (paper) mills. It stated that there were unspecified levels of hydrogen sulfide that were capable of producing nausea, vomiting, headache, loss of appetite, and disturbed sleep.

SUBACUTE POISONING

Subacute hydrogen sulfide intoxication is the rubric used to describe the clinical picture produced by the direct local irritative action of hydrogen sulfide on the moist tissues of the eyes and respiratory tract.[8, 128, 338] Signs and symptoms of subacute intoxication usually become apparent within a few hours after initial exposure to the gas. Although a brief, intense exposure may cause subacute effects, more often subacute poisoning is associated with repeated or prolonged exposure. The severity of these effects is directly related to the intensity and duration of exposure. However, above ~700 μg/liter (500 ppm), the effects of local irritation may be obscured by the more dramatic and life-threatening manifestations of acute intoxication. After recovering from the crisis of acute poisoning, many patients exhibit local irritative phenomena that require further medical attention.

The most common injury associated with subacute hydrogen sulfide poisoning results from its irritative action on eye tissues. Symptomatic irritation of the mucous membranes of the respiratory tract is somewhat less common, but under certain circumstances it may lead to extremely serious, or even fatal, complications. In this chapter these two aspects of subacute hydrogen sulfide poisoning are discussed under separate headings. It should be understood, however, that victims of subacute hydrogen sulfide poisoning usually suffer, at least to some degree, *both* eye and respiratory tract irritation.

Effects of Hydrogen Sulfide on the Eyes

Early investigators found that exposure to hydrogen sulfide at concentrations as low as 70 μg/liter (50 ppm) for approximately 1 hr produced irritation and inflammation of the conjunctival and corneal tissues, a condition they called "gas eye."[220, 338] They reported that this effect is more likely to occur under conditions of increased humidity. Because hydrogen sulfide exerts an anesthetic effect on the nerves that supply the corneal membranes, pain may not always be counted upon to provide an early warning of exposure.[260] Later, however, acute conjunctivitis develops, with characteristic signs and symptoms, including pain, lacrimation, hyperemia, retroorbital aching, blepharospasm, blurred vision, photophobia, and the illusion of rainbowlike colors around incandescent light sources.[24, 182] In it more severe form, gas eye may progress to acute keratoconjunctivitis with vesiculation of the corneal epithelium,[236] corneal ulceration, and, rarely, scar formation with permanent impairment of vision.[24] The following brief accounts are instructive in that they provide both historic and contemporary views of this most common response to hydrogen sulfide exposure.

Carson[62] stated that the earliest reference to the effects of hydrogen sulfide on the eye was made by Ramazzini[254] who, in the year 1700, noted that cesspool cleaners suffered from a peculiar eye affliction that we now know to be a characteristic response to hydrogen sulfide: "The eye is inflamed and vision obscured. The only remedy is for him to return to his house and shut himself in a darkened room and stay there until the next day, bathing his eyes . . . with tepid water."

Mitchell and Davenport,[220] in their 1924 review of hydrogen sulfide (Appendix II), specifically mentioned the "numerous cases of conjunctivitis" that occurred among workmen in the sulfur mines of Sicily, as reported by Oliver[240] in 1911. Reports of hydrogen sulfide-induced eye inflammation that were published in Eastern European journals during the 1920's were mentioned briefly by Carson.[62] In 1938, MacDonald[214] reported that an early symptom of subacute poisoning was the appearance of colored rings around lights, an effect he ascribed to edema of the outer cornea. In his experience, symptoms usually appeared between the second and fourth day of exposure. In connection with this, Carson cited the observation of Rankine,[257] who described cases involving effects on the eyes of English factory workers. These incidents always occurred on Wednesdays or Thursdays.

In 1939, Barthelemy[18] published an account of his 10 years of experience in the viscose rayon industry. The bulk of his report was devoted to a description of the occupational hazards associated with the production of viscose rayon. Also included, however, were some extremely valuable quantitative data on the relationship between air concentrations of hydrogen sulfide and carbon disulfide and the occurrence of conjunctivitis among operators engaged in spinning and washing rayon yarn. The plant described by Barthelemy used data derived from French and Italian experience to establish air standards for carbon disulfide at <0.20 mg/liter (62.6 ppm) and for hydrogen sulfide at <0.10 mg/liter (71.9 ppm).* Barthelemy noted that regular analyses from 1929 to 1933 showed that concentrations of both gases in the spinning room air were "far below" these standards. In December 1933, a period of heavy production overtaxed the capacity of the ventilating equipment and a number of spinning room operators were obliged to quit working. In accordance with findings after earlier intermittent mishaps that had demonstrated the pernicious nature of carbon disulfide intoxication but had not shown eye irritation to be an effect, the plant physician judged that none of the workers exhibited symptoms of carbon disulfide poisoning. Barthelemy did report, however, that each victim suffered severe effects involving the eye. He describes them as:

> . . . intense photophobia, spasm of the lids, excessive tearing, intense congestion, pain, blurred vision, the pupils were contracted and reacted sluggishly, the cornea was hazy and sometimes numerous blisters could be seen. The acute symptoms subsided rapidly and the corneal epithelium regenerated without scarring.

In all, 332 cases of eye injury of the type described were recorded during the month of December 1933. The average air analysis for this month was 41 μg/liter (29.5 ppm). During the next year, the ventilation system was improved. In December 1934, only 85 eye injury cases were recorded. Table 5-2 summarizes the experience in this plant over 6-year period.

Barthelemy interpreted these data as demonstrating that eye injury can be avoided if air concentrations of carbon disulfide and hydrogen sulfide are kept below 42 μg/liter (30 ppm) and 28 μg/liter (20 ppm), respectively. He emphasized his belief that carbon disulfide promotes a "hypersensitiveness" of the conjunctiva and cornea to the irritative effects of hydrogen sulfide.

A later report of hydrogen sulfide-related eye injuries in the viscose rayon industry was published by Masure.[209] His clinical description of conjunctival and corneal irritation and inflammation was in no way different than that reported by Barthelemy. A few cases resulted from exposure to concentrations of hydrogen sulfide in air as low as 5 μg/liter (3.6 ppm) to 20 μg/liter (14.4 ppm). At 30 μg/liter (21.6 ppm), this number increased considerably. The author noted that an increase in humidity appeared to favor the occurrence of symptoms.

---

* 1 mg/liter of carbon disulfide = 313 ppm at 25 C and 760 mm; 1 mg/liter of hydrogen sulfide = 719 ppm at 25 C and 760 mm.

Table 5-2. Cases of conjunctivitis in vicose rayon spinning room during the months of December from 1932 to 1937 and the average air concentrations of carbon disulfide and hydrogen sulfide[a]

Year	Cases reported, including recurrent	Average air analysis: Carbon disulfide mg/liter	ppm	Hydrogen sulfide mg/liter	ppm	Comment
1932	None	0.066	20.7	0.012	8.6	
1933	332	0.162	50.7	0.041	29.5	Increased production
1934	85	0.122	38.2	0.032	23.0	
1935	None	0.063	19.8	0.019	13.7	
1936	71	0.120	37.6	0.032	23.0	Increased production
1937	None	0.103	32.2	0.025	18.0	

[a] From Barthelemy, 1939,[18] who made no mention of the number of workers exposed.

It is probably significant that Barthelemy and Masure, each reporting independently on his experience in the viscose rayon industry, emphasized the very striking occurrence of eye injury following exposure to relatively low concentrations of hydrogen sulfide in air. From their observations, one cannot exclude the possibility that, as suggested by Barthelemy, carbon disulfide in some way enhances the effect of hydrogen sulfide of the eye.

In 1944, Larsen[182] published a report of subacute hydrogen sulfide poisoning among 50 workers engaged in constructing a tunnel under the Sund, the strait between Denmark and Sweden. Altogether, a total of 163 attacks of acute keratitis were recorded. Signs and symptoms included severe pain, photophobia, and lacrimation, combined frequently with the development of corneal vesicles that ruptured within 24 hr. Larsen noted that the men did not usually show symptoms of poisoning while actually at work in the tunnel, but, rather, after they had come out into the daylight, to which they reacted with severe photophobia and spasm of the lids. This observation led Larsen to conclude that the relatively high concentration of hydrogen sulfide in the tunnel atmosphere had a slight anaesthetic action on the nerves of the eye.

In 1951, Ahlborg[8] stated that cases of eye irritation were fairly common among Swedish shale-oil plant workers. Many of these cases resulted from accidents involving direct contact with suddenly released jets of gas. Twelve to 24 hr after such a brief but intense exposure, keratoconjunctivitis would often develop, with symptoms of pain, itching, and photophobia. Ahlborg considered secondary infection to be the most important cause of keratoconjunctivitis in such cases. He demonstrated that the number of sick leaves due to eye injury fell from 243 in 1945 to 73 in 1946. This resulted primarily from the warnings to workmen not to rub their eyes, but to rinse them with boric acid solution after a gas exposure. Ahlborg speculated that the low frequency of eye injury after acute hydrogen sulfide intoxications was due to the incapacitation that prevented the victim from contaminating his inflamed conjunctivas with his hands. On the average, eye lesions healed in 4 days. Repeated episodes did not seem to cause chronic eye damage. Nevertheless, Ahlborg suggested that persons with chronic eye inflammation not be employed in areas of hydrogen sulfide exposure.

In 1954, Nyman[236] discussed his experiences in treating victims of "gas eye" in a viscose yarn factory in Finland. He found acute conjunctivitis to be the most frequent eye disorder, with some workers developing keratitis. He vividly described the excruciatingly painful condition of ruptured corneal vesicles and the aggravation of this condition brought about by the attempted self-treatment in Finnish steam baths. The "disease picture is highly dramatic." The victim is unable to open his eyes and is afraid that he has gone blind. Nyman identified three distinct categories of hydrogen sulfide-related eye damage, which he believes represent increasing concentrations and/or duration of exposure to hydrogen sulfide: conjunctivitis alone, conjunctivitis progressing to keratitis, and immediate keratitis. After a therapeutic trial with locally administered cortisone, the second category disappeared entirely; no case of conjunctivitis progressed to keratitis. Moreover, keratitis treated with cortisone remained mild; the corneal epithelium remained intact in all his patients. The author emphasized that his new treatment saved many man-days.

Effects on the Respiratory Tract

Hydrogen sulfide is irritating to all of the mucous membranes of the respiratory tract. Rhinitis, pharyngitis, laryngitis, and bronchitis are sometimes mentioned as subacute manifestations of poisoning,[8, 269, 338] but do not appear to be very important aspects of the problem. However, the irritating effects of the gas on the deepest structures of the lung may cause severe, even fatal, pulmonary edema.[168, 210] Among the cases described by McCabe and Clayton[210] in their report of the Poza Rica, Mexico incident, nine of 47 hospitalized patients suffered from pulmonary edema, two of whom died. It is difficult and not particularly useful to categorize pulmonary edema as an acute or subacute manifestation of hydrogen sulfide poisoning. The important point is that pulmonary edema can follow clear-cut acute poisoning, or can be caused by exposures of an hour or so to concentrations of hydrogen sulfide too low to precipitate acute collapse.[280]

CHRONIC POISONING

Not all authors agree that a pathologic entity identifiable as chronic hydrogen sulfide poisoning exists. Ahlborg[8] has probably given more systematic attention to this question than any other investigator. It is his opinion that, if such a condition exists, it would be characterized as:

> a lingering poisoning, conditioned by the action over a long period, or repeatedly, of concentrations [of H_2S] which in themselves would not occasion symptoms of acute or subacute poisoning. The symptoms would probably be like those of the residual conditions found after acute poisoning, and therefore one expects to find mainly neurasthenic or otoneurological symptoms.

To clarify the question of the existence of chronic hydrogen sulfide poisoning, which is listed in The Swedish Workman's Compensation Act, Ahlborg conducted a series of studies among Swedish shale-oil industry workers. He compared two groups of workers—one consisting of 459 men exposed to the gas daily, the other of 384 men exposed only rarely. Both clinical and questionnaire data were available to the investigators. Ahlborg summarized the results of his work as follows:

> Positive proof of the existence of so-called chronic hydrogen sulfide poisoning is lacking, but examination of a great number of workers within the industry has shown that the frequency of neurasthenic troubles increases with the degree of hydrogen sulphide exposure and the length of employment.

Ahlborg cautioned against overinterpretation of the results of his study, since neurasthenia is a subjective diagnosis, in this case based largely on complaints of fatigue. Also, the author did not adjust his findings for age, a factor almost certain to be of importance in accounting for an increased frequency of fatigue and clearly related to increased length of employment.

Another study of chronic, low-level exposure to hydrogen sulfide was published by Rubin and Arieff[260] in 1945. They reported the results of a questionnaire survey of 100 workers in a viscose rayon plant where exposures to both carbon disulfide and hydrogen sulfide were well documented. Thirty-two percent of the group complained of fatigue at the end of the day's work. However, in a control group of 55 workers unexposed to either hydrogen sulfide or carbon disulfide, 34.5% complained of similar fatigue. The authors also found that shift work, rather than exposure to hydrogen sulfide or carbon disulfide, showed the highest correlation with subjective complaints, such as fatigue. The authors concluded that if chronic effects of low-grade exposure to carbon disulfide and hydrogen sulfide do, in fact, exist, they are minimal in nature.

6

Effects on Vegetation and Aquatic Animals

Until recently, hydrogen sulfide has not been considered to be an important pollutant to vegetation. Its production by industrial sources such as paper mills has certainly created an odor problem, but subsequent agricultural effects have not been deemed important until now.

Analyses of leaves near paper mills shows elevated amounts of sulfate. This is a characteristic result of hydrogen sulfide uptake and metabolism. Tree growth in these localities is probably affected by the hydrogen sulfide and, perhaps, other sulfides in the atmosphere.

The recently developed geothermal energy sources of hydrogen sulfide must also be considered. In the vicinity of geothermal wells, sulfide concentrations are usually in the range of 10 to 30 ppb, but may be as high as 100 ppb. In the vicinity of pulp mills, concentrations of sulfide will reach peaks of 100 ppb, but this includes organic sulfides as well as hydrogen sulfide. The thousands of kilograms of hydrogen sulfide that may be emitted in these operations will certainly be taken up by the surrounding vegetation. A station producing 50 MW of power a year could produce approximately 2,000 kg of hydrogen sulfide per day. This gas is a potential hazard to plants because of its contributions to both air and water pollution. Geothermal energy sources near natural vegetation may have little economic impact, except on lumber production. There may also be less tangible effects on the aesthetic and recreational value of the area, since sulfate injury to vegetation, such as growth reduction, may not be immediately apparent to the naked eye. Potential damage to vegetation and the resulting economic impact of geothermal sources should be determined before tapping such sources near agricultural areas.

As far as aquatic species are concerned, the most important sources of sulfide are from paper mill effluents and from bacterial action in the bottoms of seas, lakes, and rivers. The introduction of sulfide from the atmosphere is probably negligible. Since natural sources of sulfide have been a feature of the environment during the evolutionary process, the aqueous environment may contain species that are tolerant to low concentrations of sulfide.

INTERACTION WITH BIOLOGIC SYSTEMS

Jacques[151] studied the entrance of total sulfide into cells of the unicellular alga *Valonia macrophysa* Kütz. Penetration depended on the amount of undissociated hydrogen sulfide in the medium. Therefore, for a given concentration of total sulfide, penetration is greater at the lower pH. The studies of Bonn and Follis[40] on the survival of channel catfish (*Ictalurus punctatus* Rafin) in the acid lakes of northeast Texas indicate that this principle holds true. However, Nakamura[230] has noted effects of sulfide on algae at pH 9, perhaps indicating penetration of un-ionized hydrogen sulfide in these species.

The solubility of hydrogen sulfide suggests that the gas will be efficiently taken up by the lungs and, to some extent, by the upper respiratory tract. However, after breathing hydrogen sulfide, both animals and humans suffer from edema and other symptoms, a finding that indicates that significant quantities of the hydrogen sulfide reach the alveoli.

The uptake of gases by plants is also roughly proportional to water solubility, unless there is rapid conversion of the dissolved gas to other products.[140] The process in plants is greatly different from that in animals with lungs or gills. Plants can close their stomata, thereby curtailing the gas exchange and preventing the gas from reaching the cell surface in the interstitial leaf spaces. The cell surface available for exchange within the leaf can be from 6.8 [in the lilac (*Syringa vulgaris* L.)] to 31.3 [in the bluegum eucalyptus (*Eucalyptus globulus* Labill.)] times larger than the external surface.[314] The gas exchange by the external leaf surface, which is frequently covered by cuticle, is negligible. The toxicity of hydrogen sulfide and other pollutants is, to some extent, modified by the conditions that cause stomatal opening, e.g., illumination, water stress, and perhaps a response to the pollutant itself.

DOSE RESPONSES

Higher Plants

In 1936, McCallan *et al.*[211] noted that there was a scarcity of information on the toxicity of hydrogen sufide to plants. They surveyed 29 plant species for susceptibility by exposing potted plants in a fumigation chamber. Damage to the leaves was greatest in young, growing tissue. Leaves that were wilted immediately after the exposure became necrotic within 24 hr. The temperature during the tests varied from ~23 to 27 C. The relative humidity (RH) ranged from 82% to 100%. Hydrogen sulfide concentrations ranged from 28 to 560 μg/liter (20 to 400 ppm), and the duration of exposure was 5 hr.

Benedict and Breen[26] also measured effects of hydrogen sulfide (100 to 500 ppm) on several species. They noted considerable species differences and greater susceptibility in younger tissue.

The plant responses in these tests were variable. Resistant plants such as strawberry (*Fragaria vesca* L.) and peach (*Prunus persica* L.)

showed no damage at 280 to 560 μg/liter (200 to 400 ppm). Susceptible plants, such as cucumber (*Cucumis sativus* L.), tomato (*Lycopersicon esculentum* Mill.), and radish (*Raphanus sativus* L.), were injured at 28 to 84 μg/liter (20 to 60 ppm). The damage increased as the temperature was raised. There were some indications that wilted plants resisted injury, presumably because of stomatal closure.

Sequels to the above work were published in 1940.[19, 212, 213, 311] McCallan and Setterstrom[212] compared the lengths of the times required to kill 50% of the organisms in sets of fungi, bacteria, seeds, leaves, and stems of higher plants, and animals exposed to 1,400 μg/liter (1,000 ppm) of hydrogen cyanide, hydrogen sulfide, ammonia, chlorine, and sulfur dioxide. Seeds of rye (*Secale cereale* L.) and radish (*R. sativus* L.), both wet and dry, were not affected by hydrogen sulfide. However, the plants tested—tomato (*L. esculentum*), tobacco (*Nicotiana tabacum* L.), and buckwheat (*Fagopyrum esculentum* Moench.)—were more resistant to hydrogen sulfide than to any of the other gases tested. The animals tested (mice, rats, and houseflies) were more susceptible than the higher plants.

The results obtained by McCallan and Weedon[213] demonstrated that fungi were not particularly susceptible to hydrogen sulfide. The fungi tested were certainly more resistant than the higher plants.[212]

Thornton and Setterstrom[311] endorsed the previous finding[212] that the foliage of higher plants is more resistant to hydrogen sulfide than to hydrogen cyanide, ammonia, chlorine, and sulfur dioxide. In contrast to sulfur dioxide and chlorine, which cause an acidification of the exposed plant tissue, and ammonia, which causes an alkalinization, hydrogen sulfide had no effect on the pH of the exposed tissue.

Seed germination of *S. cereale* and *R. sativus* was not affected by hydrogen sulfide concentrations of 350 μg/liter (250 ppm) and 1,400 μg/liter (1,000 ppm) in tests made by Barton.[19] Although the results are not strictly comparable, Reynolds[258] reported similar results indicating that aqueous solutions of sulfide at 25 and 50 mg/liter have no effect on the germination of lettuce seed (*Lactuca sativa* L. cv. Arctic King).

The results reported by Dobrovolsky and Strikha[86] concerning the effects of hydrogen sulfide exposure on seed germination are in marked contrast to the two previous reports. They found that the germination of *R. sativus* seeds was inhibited when they were exposed to hydrogen sulfide concentrations as low as 0.01 μg/liter (0.0066 ppm). At 1 μg/liter (0.66 ppm), however, the germination was only 45% of control. Inhibition of both the production and size of shoots was also observed at the same hydrogen sulfide concentrations.

Faller[107] reported results of experiments in which young sunflowers (*Helianthus annuus* L.) were exposed to hydrogen sulfide fumigation, while these plants had no alternative nutrient source of sulfur. The experiment lasted 3 weeks. During this time the hydrogen sulfide gas concentration varied between a few micrograms per liter and 280 μg/liter (200 ppm). Both fresh and dry weights of the buds, the first five leaves, the stems, and the roots were taken. The plants exposed to hydrogen sulfide were heavier in all respects than the controls, which were not supplied with sulfur in the nutrient solution. There was no evidence of lesions in the exposed plants, which behaved as plants normally do when sulfate is present in the nutrient solution. Sulfur analysis of the plants showed the accumulation of sulfur, particularly in the roots. This contrasts with exposures to sulfur dioxide, during which sulfate accumulates in the leaves. These results of Faller demonstrated that hydrogen sulfide can act as the sole sulfur source for the nutrition of *H. annuus*.

McCallan *et al.*[211] reported that normally grown flowers of *H. annuus* were moderately damaged by exposure to hydrogen sulfide at 84 to 112 μg/liter (60 to 80 ppm) and 280 to 560 μg/liter (200 to 400 ppm). Thus, *H. annuus* fell into the intermediate category as far as susceptibility to hydrogen sulfide was concerned. The results of Faller[107] and McCallen *et al.*[211] are, therefore, in marked contrast.

One may conclude from these data that plants are not particularly sensitive to hydrogen sulfide at high concentrations for short periods. A study presently in progress is examining the effects of long-term exposures at low concentrations (C. R. Thompson, 1977, personal communication). Continuous exposures of alfalfa (*Medicago sativa* L.) showed that 4.2 μg/liter (3 ppm) of hydrogen sulfide caused visible lesions in 5 days. Yield was also reduced both at that concentration and, in one variety, at 0.42 μg/liter (0.3 ppm). No effect was seen at 0.042 μg/liter (0.03 ppm). In normal agricultural practice, *M. sativa* is cut, then allowed to regrow. Thompson's exposure of *M. sativa* to hydrogen sulfide resulted in successive harvests that showed yield reduction clearly at 0.42 μg/liter (0.3 ppm) for both varieties tested. Seedless grapes (*Vitis vinifera* L.) exposed to hydrogen sulfide exhibited severe damage at 4.2 μg/liter (3 ppm) and easily detectable damage at 0.42 μg/liter (0.3 ppm). White or yellow lesions were the first visible damage to leaves. Ponderosa pine (*Pinus ponderosa* Dougl. ex Lawson) showed no effect of hydrogen sulfide concentrations at 0.042 μg/liter (0.03 ppm), but developed tip burn after 8 weeks of exposure to 0.42 μg/liter (0.3 ppm). The resistance of *P. ponderosa* was consistent with the low accumulation of sulfur in the foliage.

Surprising results were obtained with *L. sativa* and sugar beets (*Beta vulgaris* L.). The yield of these vegetables increased after exposures to hydrogen sulfide at 0.042 μg/liter (0.03 ppm). These results are consistent with the results of Faller,[107] except that they were obtained at lower concentrations of hydrogen sulfide. The yield of *L. sativa* was reduced at hydrogen sulfide concentrations of 0.42 μg/liter (0.3 ppm).

Sulfide toxicity in plants can occur in waterlogged soils. Ford[111] reported this problem in citrus trees in the poorly drained flatwood areas in Florida. He determined by laboratory experiment that the threshold concentration of sulfide for root injury is 2.8 mg/liter (aqueous) after 5 days' exposure. The formation of hydrogen sulfide in these waterlogged areas can be attributed to bacterial metabolism.

Rice (*Oryza sativa* L.) also exhibits injury after exposure to sulfide. Hollis and his coworkers[11, 157, 158, 252] studied this subject in the United states. There has also been extensive research in Japan[16] and in India.[302] Joshi *et al.*[158] measured the effect of various sulfide concentrations on rice seedlings. Oxygen release, nutrient uptake, and phosphate uptake were all markedly inhibited by 1 mg/liter of sulfide, but the nutrient uptake by some varieties was stimulated by 0.05 mg/liter of sulfide. The relationship of oxygen release and nutrient uptake to resistance to physiologic responses (e.g., the conditions known as "Straighthead" and "Akagare") was studied in 28 varieties of rice. Resistant cultivars had higher oxygen release and lower nutrient uptake. The toxic effect of sulfide on rice roots is prevented if the bacterium *Beggiatoa* is present in the soil. Joshi and Hollis[157] and Pitts *et al.*[252] noted that the relationship between the rice seedlings and *Beggiatoa* is mutually beneficial. The bacteria oxidize the toxic sulfide, while the presence of the rice seedlings increases the survival of the *Beggiatoa*.

The effect of sulfide on the respiration of rice roots has been studied in some detail.[11] Respiration is inhibited 14% by 0.1 mg/liter (aqueous) of sulfide, and 25.6% by 3.2 mg/liter (aqueous). Homogenates were prepared from rice roots that had been exposed for 3 to 6 hr to 0.1 to 3.2 mg/liter (aqueous) of sulfide. Assays were then made of various oxidase enzymes. Ascorbic acid oxidase, polyphenol oxidase (both copper-containing), and catalase, peroxidase and cytochrome *c* oxidase (all heme-containing), were inhibited in homogenates from all levels of sulfide treatment. The most striking inhibition occurred with cytochrome oxidase, of which 40% was inhibited after a 6-hr pretreatment with 0.1 mg/liter of sulfide.[11]

Algae

A comprehensive article on the effects of hydrogen sulfide on algae was written by Czurda,[80] who dissolved sulfide in the nutrient solutions. The results of Czurda's experiments bear more directly on the effects of bacteria-generated sulfides in aqueous environments rather than on anthropogenic hydrogen sulfide in the ambient air. Czurda found that some species and strains of algae could multiply in sulfide concentrations of 8 to 16 mg/liter (aqueous), whereas others were inhibited at concentrations of 1 to 2 mg/liter (aqueous).[80] He pointed out that one cannot speak of hydrogen sulfide resistance in general when discussing algae, for there are differential effects on cell division, respiration, assimilation, and fermentative ability.

Nakamura[230] confined his attention to two algae in his study: *Pinnularia* sp. and *Oscillatoria* sp. In the presence of 32 mg/liter (aqueous) of sulfide, the number of colonies of *Pinnularia* approximately doubled, but the number of colonies of *Oscillatoria* diminished ~15%. In agreement with Czurda,[80] Nakamura found that metabolic parameters were variously affected by sulfide (see "Plants" under "Physiologic Studies" on p. 76).

Aquatic Species

Marine Theede *et al.*[306] provided a summary of information concerning effects of hydrogen sulfide on marine organisms. They reported that sulfide concentrations of 7 mg/liter were found in the Black Sea at depths below 2,000 m, and that 6.13 mg/liter of sulfide was recorded in the North Sea mudflats. In their own experiments, the exposed lamellibranchs (pelycyopods), gastropods, polychaetes, crustaceans, and echinoderms to sulfide concentrations of approximately 7.5 mg/liter. They observed pronounced differences among the species of ciliary activity and survival capacity of isolated gill tissue. The effects of sulfide were less pronounced at colder temperatures, and with mussels (*Mytilus edulus* L.), gill tissue survival was better at pH 7 than at pH 8.

Fresh water There have been several reports of effects of sulfide on fresh-water species. Colby and Smith[71] made a comprehensive study of the effects of paper mill effluents. They measured sulfide concentrations in the water at several depths and at various distances downstream from paper mills. As far as 99 km from the paper mills, they observed oxygen deficiencies and elevated sulfide concentrations near the interface of water and sludge deposits. They thoroughly analyzed the pH and the dissolved oxygen and sulfide concentrations at different depths and temperatures. These parameters were compared with the survival of eggs from walleyed pike (*Stizostedion vitreum vitreum* Mitch.). Eggs placed on mats 30 cm above the bottom survived better than those placed on the bottom. In the laboratory, under conditions approximating those on the river, sulfide levels of 0.3 mg/liter (aqueous) were lethal to gammarids (*Gammarus pseudolimnaeus* Horsfield) and to *S. vitreum vitreum* eggs and fry. Sensitivity to sulfides was greater at lower concentrations of dissolved oxygen. The concentrations of dissolved sulfides found in the river water ranged up to 8 mg/liter (aqueous).

Adelman and Smith[6] made a systematic study of the interrelationship of sulfide toxicity and oxygen concentration. They determined the mean tolerance limits (TL_m) to sulfide for the eggs and fry of northern pike (*Esox lucius* L.) at oxygen concentrations of 2 ppm and 6 ppm (see Table 6-1). They reported that the eggs are more resistant than the fry,

Table 6-1. Sulfide TL_m values for the eggs and fry of *Esox lucius* L.[a]

	Sulfide [mg/liter (aqueous)]			
	Eggs		Fry	
Oxygen (ppm)	TL_m 48 hr	TL_m 96 hr	TL_m 48 hr	TL_m 96 hr
2	0.076	0.034	0.016	0.009
6	0.046	0.037	0.047	0.026

[a] Data condensed from Adelman and Smith, 1970.[6]

the maximum safe sulfide concentration being 0.014 to 0.018 mg/liter (aqueous) for eggs and 0.005 to 0.006 mg/liter (aqueous) for fry. The ameliorative effect of higher oxygen concentration is more apparent in the fry than in the eggs.

The chronic toxicity of sulfide on *G. pseudolimnaeus* was studied by Oseid and Smith.[241] A preliminary study showed that the LC_{50} was 0.022 mg/liter in an experiment lasting 96 hr. However, tests run for 65, 95, and 105 days showed that the maximum safe level of sulfide was 0.002 mg/liter: 10 times less than the 96-hr LC_{50}.

The poor yield of *I. punctatus* in stocked acid lakes is attributable to lethal amounts of sulfide. This problem can be solved by stocking the lakes with young adult fish, which are relatively resistant to sulfide, or by raising the pH of the lakes. The beneficial effect of the higher pH results because the toxicity of sulfide is less than that of undissociated hydrogen sulfide. Bonn and Follis[40] reported that the TL_m of un-ionized hydrogen sulfide was 0.8 mg/liter (aqueous) at pH 6.8 and 0.53 mg/liter (aqueous) at pH 7.8. At pH 6.8, undissociated hydrogen sulfide was about 50% of the total sulfide; at pH 7.8, it is about 10% of total sulfide. Bonn and Follis[40] noted that in the shallow acid lakes of northeast Texas hydrogen sulfide reached its minimum concentration (0.15 mg/liter) in the winter months, and rose to its highest concentrations (0.8 mg/liter) of un-ionized hydrogen in the spring, presumably because the increased temperatures favored the bacterial production of sulfide.

The Environmental Studies Board of the National Academy of Engineering[231] reported the maximum safe concentration of undissociated hydrogen sulfide to be 2 μg/liter. Furthermore, they suggested, to protect aquatic organisms, 2 μg/liter of total sulfides should not be exceeded.

METABOLISM

Enzymologic Studies

Cysteine synthase [*O*-acetyl-L-serine acetate-lyase (adding hydrogen sulfide)] (EC 4.2.99.8). This enzyme catalyzes the reaction:

$$\underset{O\text{-acetyl-L-serine}}{CH_3CO{-}OCH_2\overset{\overset{\displaystyle NH_2}{|}}{C}HCOOH} + \underset{\text{hydrogen sulfide}}{H_2S} \xrightarrow{\text{(cysteine synthase)}} \underset{\text{cysteine}}{HSCH_2\overset{\overset{\displaystyle NH_2}{|}}{C}HCOOH} + \underset{\text{acetate}}{CH_3COO^-} + \underset{\text{hydrogen}}{H^+} \qquad (1)$$

Thompson and Moore[309] reported on the cysteine synthase in bread mold (*Neurospora crassa* Shear and Dodge), yeast (*Saccharomyces cerevisiae* Meyer ex. Hansen), and in turnip leaves (*Brassica rapa* L.). *O*-Acetylserine was a hundredfold more active than serine and is considered to be the naturally occurring reactant. The reactions were conducted at pH 8.0 and at sulfide concentrations of 69 mg/liter. In yeast extracts, formation of methyl cysteine from methyl mercaptan was better than cysteine formed from sulfide, but in turnip leaf preparations cysteine formation was about twice the amount of methyl cysteine formation.

Ngo and Shargool[233] studied substrate specificity of cysteine synthase from germinating seeds of rape (*Brassica napus* L.). They found that the sulfide consumed by serine, phosphoserine, *O*-acetylhomoserine, and *O*-succinylhomoserine was 6.3%, 10.4%, 4.4%, and 2.8%, respectively, of that consumed by *O*-acetylserine. Ngo and Shargool[232] also determined kinetic parameters for the cysteine synthase from germinating seeds of *B. napus*. They found a K_m (Michaelis constant) for *O*-acetylserine of 1.7 μM and a K_m of 0.43 mM for sulfide (13.76 mg/liter of sulfide). The concentration of sulfide required for half-maximal velocity is thereafter relatively high. It is not certain that sulfide is the natural donor for cysteine biosynthesis. The predominant source of sulfide under natural conditions would be the reduction of sulfate. It is possible that an intermediate of reduction rather than free sulfide is the actual donor.

Frankhauser *et al.*[108] found that 20% of cysteine synthase is localized in the chloroplasts of spinach (*Spinacia oleracea* L.).

Methionine synthase [*O*-acetyl-L-homoserine acetate-lyase (adding methanethiol)] (EC 4.2.99.10) This enzyme catalyzes the reactions:

$$\underset{O\text{-acetyl-L-homoserine}}{CH_3CO{-}O(CH_2)_2\overset{\overset{\displaystyle NH_2}{|}}{C}HCOOH} + \underset{\text{methanethiol}}{CH_3SH} \xrightarrow{\text{(Methionine synthase)}} \underset{\text{methionine}}{CH_3S(CH_2)_2\overset{\overset{\displaystyle NH_2}{|}}{C}HCOOH} + \underset{\text{acetate}}{CH_3COO^-} + \underset{\text{hydrogen}}{H^+} \qquad (2)$$

and:

$$\underset{O\text{-acetyl-L-homoserine}}{CH_3CO{-}O(CH_2)_2\overset{NH_2}{\overset{|}{C}}HCOOH} + \underset{\text{hydrogen sulfide}}{H_2S} \xrightarrow{\text{(Methionine synthase)}} \underset{\text{homocysteine}}{HS(CH_2)_2\overset{NH_2}{\overset{|}{C}}HCOOH} + \underset{\text{acetate}}{CH_3COO^-} + \underset{\text{hydrogen}}{H^+} \quad (3)$$

Giovanelli and Mudd[119] used extracts from spinach leaves to show that these reactions with *O*-acetylhomoserine could be clearly resolved from reactions using *O*-acetylserine by ammonium sulfate fractionation. Formation of methionine was almost twice as effective as formation of homocysteine.

β-Cyanoalanine synthase [L-cysteine hydrogen-sulfide-lyase (adding hydrogen cyanide)] (EC 4.4.1.9) This enzyme catalyzes the reaction:

$$\underset{\text{L-cysteine}}{HSCH_2\overset{NH_2}{\overset{|}{C}}HCOOH} + \underset{\text{hydrogen cyanide}}{HCN} \xrightarrow{(\beta\text{-cyanoalanine synthase})} \underset{\text{3-cyanoalanine}}{NCCH_2\overset{NH_2}{\overset{|}{C}}HCOOH} + \underset{\text{hydrogen sulfide}}{H_2S} \quad (4)$$

Hendrickson and Conn[134] studied this enzyme obtained from the seeds of blue lupine (*Lupinus angustifolia* L.). The purified enzyme can also synthesize β-cyanoalanine from *O*-acetylserine and hydrogen cyanide but at a little better than one-twentieth of the rate with cysteine. Conversely, cysteine synthase can catalyze the formation of β-cyanoalanine from *O*-acetylserine and hyrogen cyanide, but at only one-tenth of the rate for cysteine synthesis.

The first two enzymes listed above are potentially capable of utilizing sulfide taken up by the plants. In actuality, this is probably a minor pathway of sulfide metabolism.

Schnyder and Erismann[275] noted that several sulfur-containing amino acids were labeled after exposure of pea seedlings to $H_2{}^{35}S$. One of these compounds was found to be identical to thiothreonine (α-amino-β-thiobutyric acid).[275] A later paper by Schnyder *et al.*[276] confirmed the identify of the thiothreonine and demonstrated that its formation in pea seedling homogenates was stimulated more by added phosphohomoserine than by phosphothreonine.[276] They also reported that the synthesis of threonine in extracts from pea seedlings (*Pisum sativum* L.) and *Lemna* sp. was inhibited by sulfide.[276] Neither the mechanism of inhibition nor the pathway of incorporation of sulfide into thiothreonine has been established. Schnyder *et al.*[276] concluded that incorporation of sulfide into thiothreonine is part of a scavenging mechanism for toxic concentrations of sulfide.

Physiologic Studies

Animals: background for aquatic species There is no information on the metabolism of sulfide in fish (see Chapter 5 for discussion of sulfide metabolism in other animals). Even though it is not clear whether sulfide oxidation is enzymic, it is clear that oxidation is a major fate of sulfide. These results have not yet been verified or contradicted for aquatic animals.

Plants Brunold and Erismann[48] made a thorough study of the floating water plant *Lemna minor* L. when it was exposed to hydrogen sulfide. Metabolism of sulfide was compared to that of sulfate. Analytical results are listed in Table 6-2. These data show that sulfide is mostly converted to sulfate and that the sulfur analysis of plants supplied with both sulfide and sulfate closely replicated those supplied with sulfide alone. The exposure of *L. minor* to 25 μg/liter (18 ppm) of hydrogen sulfide caused a rapid 30% increase in cysteine content, which quickly stabilized. However, this increase amounted to only 10 μg of cysteine sulfur/g dry weight. Accumulation of sulfate was slower but linear as a function of time [$\sim$100 μg of sulfate in a 3-hr exposure to 250 μg/liter (180 ppm) of hydrogen sulfide].

Hydrogen sulfide at 25 μg/liter (18 ppm) inhibited the uptake of both phosphate and sulfate. The effect on phosphate was small and there was an immediate return to control when hydrogen sulfide gassing ceased. Sulfate uptake was inhibited 80% after a 90-min exposure to 25 μg/liter (18 ppm) of hydrogen sulfide; 100 min after the gassing ceased, recovery was to less than 50% of control. The effect of hydrogen sulfide on apparent photosynthesis was a small inhibition between 0 and 7

Table 6-2. Total sulfur, sulfate, and sulfide analysis of *Lemma minor* L.[a]

Sulfur source	Total sulfur[b]	Sulfate[b]	Sulfide[b]
0.4 mM SO_4^{2-}	2.42 ± 0.12	0.50 ± 0.02	—
6.0 ppm H_2S	6.54 ± 0.43	4.50 ± 0.17	0.032 ± 0.002
0.4 mM SO_4^{2-} + 6.0 ppm H_2S	6.7 ± 0.41	4.58 ± 0.18	0.029 ± 0.002

[a] Data condensed from Brunold and Erismann, 1974.[48]
[b] Expressed as mg/g dry weight.

μg/liter (5 ppm), but no further decrease until 84 μg/liter (60 ppm) of hydrogen sulfide was exceeded.

Brunold and Erismann[49] showed that incorporation of sulfide into cysteine in extracts of *L. minor* was by the *O*-acetylcysteine pathway. In contrast to assimilation of sulfate into cysteine, this incorporation of sulfide was not dependent on light. This indicated direct incorporation of sulfide into cysteine, rather than conversion to sulfate before incorporation. Pulse labeling with sulfide gave results supporting this conclusion, since cysteine was rapidly labeled, whereas sulfate was slowly labeled at the time the label in cysteine was decreasing.[49]

Nakamura[230] studied the effects of dissolved sulfide on two algae species, *Pinnularia* sp. and *Oscillatoria* sp. At pH 7.2 in the presence of 0.1 mM [3.2 mg/liter (aqueous)] total sulfide, catalase activity of both species was completely inhibited. However, colony formation in *Oscillatoria* was only slightly inhibited by 1 mM [32 mg/liter (aqueous)] total sulfide and in *Pinnularia* it was actually stimulated twofold. In darkness, oxygen uptake was stimulated in both species by 1 mM [32 mg/liter (aqueous)] and 0.1 mM [3.2 mg/liter (aqueous)] sulfide, the lower concentration being somewhat more stimulatory. Under photosynthetic conditions, sulfide strongly inhibited oxygen evolution, even at concentrations of 10 μM [0.32 mg/liter (aqueous)] at pH 9. On the other hand, photosynthetic carbon dioxide fixation was not markedly affected by sulfide.

The strong inhibition of catalase noted by Nakamura[230] was also observed by Dobrovolsky and Strikha[86] in their study of effects of hydrogen sulfide on germinating seeds.

While discussing metabolism of sulfide it should be mentioned that sulfite reduction produces sulfide in photosynthetic systems,[262] and in cases in which $^{35}SO_2$ has been supplied to illuminated plants, the evolution of $H_2{}^{35}S$ has been detected.[83]

MODE OF TOXICITY

Slater[282] compared the inhibition of NADH (reduced nicotinamide adenine dinucleotide) oxidation by mitochondrial preparations in the presence of various inhibitors. He reported 96.3% inhibition by 0.1 mM [3.2 mg/liter (aqueous)] sulfide at pH 7.3, while 96.9% inhibition was caused by 0.5 mM cyanide. He considered these results to be a consequence of complexation of both reagents with the heme moieties of cytochrome oxidase. Nicholls[235] has determined that the inhibition of cytochrome oxidase (cytochrome aa_3) is similar to that of cyanide in that it is slow binding and has a high affinity. It is different from cyanide in that the binding is independent of the redox state of components other than aa_3. Inhibition of catalase by sulfide[86, 230] may also be attributed to heme binding.

Gassman[117] reported that chlorophyll biosynthesis is inhibited by sulfide. Specifically, the step between protochlorophyll(ide)$_{650}$ and protochlorophyll(ide)$_{633}$ is stimulated, but the latter substance cannot be converted to the 650 wavelength if the hydrogen sulfide exposure exceeds 3 min. Cyanide and azide cause irreversible conversion to the photoinactive protochlorophyll(ide)$_{633}$, showing again the similarity of the three inhibitors. The chemical mechanism of this effect of sulfide is not understood, but it will be interesting to see if a heme compound is involved in the process.

SUMMARY

Plant species differ widely in susceptibility to hydrogen sulfide.

Long-term exposures to hydrogen sulfide show injury at concentrations between 0.042 μg/liter (0.03 ppm) and 0.42 μg/liter (0.3 ppm). At 0.042 μg/liter (0.03 ppm) some species (e.g., *L. sativa*, *B. vulgaris*) actually show growth stimulation, but they are also damaged at 0.42 μg/liter (0.3 ppm).

Most of the hydrogen sulfide taken up by plants is metabolized to sulfate.

Experiments with algae have shown that different metabolic processes are differentially susceptible to hydrogen sulfide.

The susceptibility of fish to sulfide depends markedly on pH—it increases as the acidity increases.

The safe concentration of sulfide in fresh water has been set as 0.002 mg/liter as a result of chronic toxicity tests on *G. pseudolimnaeus*.

The biochemical basis for the effects of sulfide on plants and aquatic animals is not understood.

RECOMMENDATIONS

On the basis of current studies on a variety of plant species, the maximum concentration of hydrogen sulfide at which damage can be avoided is 0.042 μg/liter (0.03 ppm).

The previously set safe concentration of sulfide in water (0.002 mg/liter) should be retained.

Further research on the intermittent exposure of plants to hydrogen sulfide should be conducted.

Dose/response relationships for damage to plants should be established to see if the response is linear throughout the dose range or if there is a threshold before damage starts.

Reserarch directed toward understanding the physiologic and biochemical bases for hydrogen sulfide toxicity should be supported.

7

Air Quality Standards

Hydrogen sulfide is a colorless gas that has an obnoxious odor at low concentrations. The odor theshold is in the micrograms per cubic meter range. In higher concentrations, the gas is toxic to humans and animals and is corrosive to many metals. It reacts with heavy metals in paints, thereby causing discoloration and tarnishes silver. In humans, at low concentrations it causes headache, conjunctivitis, sleeplessness, pain in the eyes, and similar symptoms, and at high concentrations in air, it produces complete fatigue of the olfactory nerve and, eventually, death. However, the majority of the complaints come from its obnoxious odor at low concentrations.

Air pollution by hydrogen sulfide is not a widespread urban problem. It is generally localized near emitters, such as kraft paper mills, industrial waste disposal ponds, sewage treatment plants, petroleum refineries, and coke ovens. The anticipated development of geothermal energy will create substantial additional sources of hydrogen sulfide from steam wells and geothermal power stations.

AIR QUALITY STANDARDS

The establishment of standards or criteria for hydrogen sulfide has often been difficult. To understand these standards fully, the methods used to establish them should be reviewed. Many critical decisions must be made regarding the existence or extent of the relationships between various air pollution levels and their effects. These relationships become quite confused in the interpretations of mortality and morbidity data on people, plants, and animals exposed for their lifetimes to a variety of stresses, including air pollution.

The attempt to determine if a single air pollutant in the presence of other pollutants can cause a certain physiologic or psychological problem can lead to uncertain conclusions. There is also doubt concerning the extrapolation to humans of animal data obtained from controlled exposures.

Adverse effects in receptors must be defined. The implications of damage or injury are not always evident. As experimental techniques improve, so also will the ability to detect subtle changes from physiologic and psychological norms that can be attributed to pollution. The norm in this case is exposure to unpolluted air. The deviations may be reversible when exposure to the pollutant stops. It has been argued that reversible environmental and biologic effects should not be used on the bases for a standard.[296] A safer, more prudent position would be to consider any measurable deviations from the norm as deleterious until proven benign.

Within any species there generally exists a range of resistance and susceptibility to air pollutants such as hydrogen sulfide. If the species is not human, there is the additional problem of correlating their data with those for humans. Even with humans, a range of susceptibility exists. Air pollution levels, in this case those for hydrogen sulfide, must be safe not only for the healthy adult, but also for the aged, the infirm, and infants, who, as a group, are the most susceptible to the effects of hydrogen sulfide.

Another factor to be considered is the darkening of lead-based paints caused by hydrogen sulfide. Because this effect detracts from the paint's appearance, it could be considered when setting guidelines or standards for hydrogen sulfide. Hydrogen sulfide also reacts with metals. In an industrial environment, it can affect silver, copper, and even gold. These reactions not only have aesthetic consequences, but also could cause malfunction of such equipment as computers by increasing the resistance of electric contacts.

AMBIENT VERSUS OCCUPATIONAL STANDARDS

Threshold Limit Values

A distinction must be made between air quality standards for the ambient air and threshold limit values (TLV) for workroom atmospheres. TLVs are those doses that, based on available data, cause no evident harm to most workers who are exposed 7 or 8 hr/day for 5 days/week. A small percentage of workers may experience discomfort from some substances at concentrations at or below the TLV. The TLVs should be used as guides for controlling health hazards, not for distinguishing the fine line between safe and dangerous concentrations.[12]

The American Conference on Governmental Industrial Hygienists (ACGIH), which is responsible for determining TLVs, has set the value for hydrogen sulfide at 15 mg/m^3 (10 ppm). The time-weighted averages, based on the 8-hr workday and 40-hr workweek, permit excursions above this limit in the industrial environment provided they are compensated by equivalent excursions below the limit during the workday. The degree of permissible excursion is related to the magnitude of the threshold limit value of a particular substance. The TLV for hydrogen sulfide falls between 10 and 100 ppm. According to ACGIH, this yields an excursion factor of 1.5.[12] Therefore, for hydrogen sulfide the maximum concentration permitted for a short time ($\leq$15 min) would be 15 ppm. These limiting excursions should be used only as "rule of thumb" guides. They may not always provide the most appropriate excursion for a particular substance.

Ambient Air Quality Standards

Air quality standards were formerly established by several quite different approaches. One method was used when community A said, "We will be satisfied if our air quality is as good as that in community B." Knowledge of air quality in community B thus provides a basis for the standard for community A.[296] A second approach was to select a date back in time and to say that the air quality then would satisfactorily meet present-day standards. This worked if the air quality was measured on the baseline year. If not, then the air quality on an earlier date could be computed by using past and present emission data. A third approach was to use the air quality on days of good ventilation as a standard. Various combinations of these three approaches can be used to arrive at an air quality standard, which must take into account not only air quality criteria but also the air quality and emission data that exist within the community.

EXISTING HYDROGEN SULFIDE STANDARDS

Air Quality

Air quality standards are being developed all over the world. In the United States, national air quality standards to protect human health are set on the federal level by the U.S. Environmental Protection Agency (EPA) for outdoor exposures and by the Occupational Safety and Health Agency (OSHA) for occupational exposures. State and local governments may establish stricter standards if they are justified.

The EPA has established National Primary Air Quality Standards for six pollutants in the ambient air, using the EPA air quality criteria documents as a basis.[316-320, 323] A standard for hydrogen sulfide has not yet been formulated by the EPA.

California, Missouri, Montana, New York, Pennsylvania, and Texas are among the forerunners that have developed independent regional standards for air quality. Doubtless other states will soon follow this trend. A tabulation of permissible ambient hydrogen sulfide concentrations is shown in Table 7-1.

Several countries have also shown some concern about hydrogen sulfide pollution. Their governments have already adopted hydrogen sulfide standards for ambient air quality (see Table 7-2). Japan adopted its Air

Table 7-1. Ambient air quality standard for hydrogen sulfide for the United States[a]: All standards are primary standards unless otherwise noted

Location	Long term[b]			Short term[b]			Remarks
	mg/m^3	ppm	Averaging time (hr)	mg/m^3	ppm	Averaging time (min)	
California				0.042		60	
Kentucky					0.01	60	Secondary standard
Minnesota				0.042		30	Not to be exceeded more than twice in any 5 consecutive days
				0.07		30	Not to be exceeded more than twice a year
Montana				0.042		30	Not to be exceeded more than twice in any 5 consecutive days
				0.07		30	Not to be exceeded more than twice a year
New Mexico				0.0042		30	Only hydrogen sulfide
				0.0042		30	Total reduced sulfur
New York				0.014		30	
North Dakota				0.045		30	Not to be exceeded more than twice in any 5 consecutive days

Oklahoma						
Tulsa City				0.03	30	Not to be exceeded more than once in any 5 consecutive days
Tulsa County				0.05	30	24-hr average, not to be exceeded more than once a year
Pennsylvania	0.005	24		0.1	60	
Tennessee						
Nashville				0.03	30	Not to be exceeded more than once in any 5 consecutive days
Davidson County				0.05	30	24-hr average, not to be exceeded more than once a year
Texas						
Residential and recreational area			0.112		30	
Industrial area			0.168		30	
Wyoming			0.04		30	Not to be exceeded more than twice in any 5 consecutive days
			0.07		30	Not to be exceeded more than twice a year

[a] From Martin and Stern, 1974.[208]

[b] Long term has no other meaning than ''long averaging time'' (greater than 3 hr); short term is less than 3 hr.

Table 7-2. Ambient air quality standard for hydrogen sulfide other than those from subsidiary jurisdictions of the United States[a]

Location	Long term[b]			Short term[b]			Remarks
	mg/m³	ppm	Averaging time (hr)	mg/m³	ppm	Averaging time (min)	
Bulgaria	0.008[c]	0.005	24	0.008	0.005	20	
Canada							
Alberta	0.004	0.003	24	0.014	0.009	60	
Alberta				0.017	0.011	30	
Manitoba	0.017	0.011	24	0.028	0.018	60	
Newfoundland				0.03	0.02	60	Maximum acceptable level
Ontario				0.03	0.02	60	Criteria for desirable ambient air quality
Saskatchewan	0.007	0.005	24	0.07	0.05	60	Provisional maximum quantities, 1970
Czechoslovakia	0.008	0.005	24	0.008	0.005	30	
Democratic Republic of Germany (East Germany)	0.008	0.005	24	0.015	0.01	30	Permissible standard, averaging time is defined as 10 to 30 min
Federal Republic of Germany (West Germany)	0.15	0.1	0.5	0.3	0.2	30	Verein Deutscher Ingenieure[325] Short-term standard = short-exposure limit, not to be exceeded more than once in any 8 hr
	0.15	0.1	0.5	0.3	0.2	30	
	0.02	0.013	0.5	0.05	0.03	30	Proposed federal standard (stations of October 1973)

Finland	0.05	0.3	24	0.15	0.1	30	Not national legal norms, communal health councils can enforce them
Hungary	0.15	0.1	24	0.3	0.2	30	
	0.008	0.005	24	0.008	0.005	30	Highly protected and protected areas
Israel	0.045	0.03	24	0.15	0.1	30	National air quality standard
Italy	0.04	0.03	24	0.1	0.07	30	Not to be exceeded more than once in any 8 hr
Poland	0.02	0.013	24	0.06	0.04	20	For protection areas
	0.008	0.005	24	0.008	0.005	20	Special protection area
Romania	0.01	0.006	24	0.03	0.02	30	
Spain	0.004	0.00025	24	0.01	0.006	30	Proposed standard
Union of Soviet Socialist Republics (USSR)	0.008	0.005	24	0.008	0.005	30	If several substances with synergistic toxic properties are present in the air, then USSR uses formulas for establishing the maximum permissible concentration
Yugoslavia	0.008	0.005	24	0.008	0.005	30	

[a] From Martin and Stern, 1974.[207]

[b] The terms "short term" and "long term," if not otherwise stated, reflect only short or long averaging times.

[c] Underlined concentrations represent the values listed in legislation; others are approximate conversions.

Table 7-3. Emission standards for hydrogen sulfide in effluent air or gas from stationary sources in the United States[a]

Source	Location	Standard	Remarks
Any	California (see remarks column)	10 ppm	Six counties: Los Angeles, Orange, Riverside, San Bernardino, Santa Barbara, and Ventura
	Connecticut	10 grains/100 ft^3	
	Norwalk		Emission rate based on process weight
	Stamford		Emission rate based on process weight
	Indiana (East Chicago)	160 ppm	For 2 min
	Kansas	10 grains/100 ft^3	
	Mississippi	1 grain/100 ft^3	Or incinerate at 871.11 C for 0.5 sec
	Montana	50 grains/100 ft^3	Burning prohibited
	New Hampshire	5 grains/100 ft^3	
	Ohio	100 grains/100 ft^3	
	Oklahoma (Tulsa)	100 ppm	
	Texas		Ground level concentration not to exceed 0.08 ppm (30-min average) for residential, business, or commercial property, or 0.12 ppm (30-min average) for other property (0.3 ppm during shut-down or start-up)
	Virginia	15 grains/100 ft^3	

Burning as fuel (oil refinery)	All states	230 mg/m³	Except by flaring for pressure relief
			Federal new source performance standard
Coke ovens	New York	50 grains/100 ft³	
Flaring	Kentucky (Priority I AQCR)[b]	10 grains/100 ft³	Louisville and Cincinnati Air Quality Control Regions
	Pennsylvania	50 grains/100 ft³	
	Puerto Rico	10 grains/100 ft³	
	West Virginia	50 grains/100 ft³	
Gas plants	New Mexico	10 ppm	
		100 ppm	Combinations of carbon disulfide, hydrogen sulfide, and carbon oxysulfide
Kraft pulp mills	Oregon and Washington	17.5 ppm	Total reduced sulfur as hydrogen sulfide on a dry gas basis
Refinery process	Alabama (see remarks column)	150 ppm	Three counties: Huntsville, Jefferson, and Mobile
	Michigan (Wayne County)	100 grains/100 ft³	
Refinery process start-up	Texas	0.3 ppm	Also shut-down

[a] From Martin and Stern, 1974.[208]

[b] Air Quality Control Region in which priority area has been designated by the U.S. Environmental Protection Agency.

Table 7-4. The emission standard for hydrogen sulfide in effluent air or gas from stationary sources other than those from subsidiary jurisdictions of the United States[a]

Source	Location	Standard original units	mg/m^3	Remarks
Any	Czechoslovakia	0.08 kg/hr		Emission rate above which it is necessary to submit a report to the government. Where discharge is for <1 hr, there is a proportionate increase in emission rate permissible without such reporting
	Great Britain	5.0 ppm	7.5	
	Singapore	5.0 ppm	7.5	
Any trade, industry, or process	Australia	mg/m^3	5.0	STP at 0 C and 1 atm (dry) National guidelines for new plants As hydrogen sulfide
	New South Wales	5.0 ppm	7.5	
	Queensland	5.0 ppm	7.5	
	Victoria	5.0 ppm	7.5	
Kraft pulp mill recovery furnace	Sweden	mg/m^3	10.0	99% of the time per month for new units, 90% for existing units; also, $\frac{\text{concentration in stack gas}}{\text{concentration at odor threshold}}$ = at least 10,000

Kraft pulp mill recovery stack	Canada			
	British Columbia	5.0 ppm	7.5	Objective Level A—average value for 24-hr period
		20.0 ppm	30.0	Objective Level B—average value for 24-hr period
		70.0 ppm	105.0	Objective Level C—average value for 24-hr period
Petroleum refineries	Federal Republic of Germany (West Germany)	1 g/m³	1,000.0	Proposed federal standard (status of October 1973) If hydrogen sulfide concentration is 10% volume, gases have to be treated or burned. After treatment, limit is 2 mg hydrogen sulfide/m³
	United States	mg/m³	230.0	Proposed Unless burned to sulfur dioxide in a manner that prevents release of sulfur dioxide to atmosphere
Waste coke oven gas	Federal Republic of Germany (West Germany)	1.5 g/m³	1,500.0	Verein Deutscher Ingenieure[326] Other sulfuric compounds 500 mg/m³
Coke oven gas (hydrogen sulfide and compounds)	Federal Republic of Germany (West Germany)	1.5 g/m³	1,500.0	Proposed federal standard (status of October 1973) An hourly average, other sulfuric compound, 0.5 mg/m³

[a] From Martin and Stern, 1974.[207]

Pollution Control Law in 1972; however, this ordinance does not mention hydrogen sulfide specifically. Great Britain has two sets of regulations to control the air quality. The Works Regulations Act of 1906, revised in 1966 and 1971, covers manufacturing processes and industry. The Clean Air Acts of 1965 and 1968 contain regulations for domestic and commercial furnaces. The application of the provisions of the Clean Air Acts are largely the responsibility of local authorities. If, in their opinion, a problem does not exist, then no action is taken. France has no regulation concerning hydrogen sulfide. However, plans for new plants must be reviewed to determine whether or not they include satisfactory emission control equipment.

Emissions

Emission standards for hydrogen sulfide have also been set in various countries. These standards limit the concentration or rate at which a pollutant is emitted from a source. The concentration of a pollutant in the effluent may be discerned subjectively, by smelling its odor, or objectively, in terms of its weight or volume. Emission standards may be derived by considering air quality criteria, manufacturing process, or fuel and/or equipment. Emission standards sometimes reflect economic, sociologic, and political factors as well as technologic considerations. In many cases, available technologic capability to control specific pollutants is not being implemented because of economic, social, or political reasons. Conversely, proper motivation is often present when the technical data are not available.

Tables 7-3 and 7-4 give the emission standards for hydrogen sulfide in effluent air or gas for some jurisdictions within and outside the United States.

8

The Psychological and Aesthetic Aspects of Odor

One of the most pronounced characteristics of hydrogen sulfide is its distinctive odor, which most people find unpleasant. Because this property of the gas is so well known, the Subcommittee on Hydrogen Sulfide decided to include the following material on the psychological and aesthetic aspects of odor in general, and hydrogen sulfide in particular.

Recent research on olfaction has established the importance of pheromones, which are volatile secretions from animals that can elicit one or more behavioral responses when perceived by members of the same species. They can influence sexual activity, serve as warning signals, and delineate trails and territory.[101] The possible existence of pheromones in humans has led to a renewed interest in the sense of smell. A popular manifestation of this is the recent emphasis on aphrodisiacs in perfumes. There is also an increasing interest in the use of smell by humans to warn them of contaminated environments. Moncrieff[223] has suggested that as our environment becomes increasingly polluted, the sense of smell may become more important to humans. This is now generally recognized not only by scientists but also by the general public and by politicians.

This attitude is in sharp contrast to the concept that smell is important to animals, but to civilized humans it is only a contribution to taste. Some recent surveys in the United States show that about one-third of the complaints received by air pollution authorities from citizens were concerned with unpleasant or noxious odors, often in the absence of violation by industry.[278] Thus, the problem is not purely physical, chemical, or even medical.

This chapter emphasizes the psychological and functional aspects of the sense of smell in evaluating the air we breathe and the food we eat. A complete psychophysical understanding requires physical and chemical analysis as well. This chapter is divided into two main topics—psychophysics and aesthetics. Psychophysics is the classic study of sensory psychology, with emphasis on detection, discrimination, and adaptation, i.e., areas in which psychology overlaps with physiology.

Very little work has been devoted to the aesthetics of odor. Nonetheless, this chapter reports what is known about the preferences of people for different odor qualities. For example, most people would judge hydrogen sulfide as unpleasant and lavender as pleasant. The factors that influence this kind of judgment or reaction are discussed below. Not everyone would consider this a part of aesthetics. In this connection, scientific research on the sense and, of course, any study of aesthetics, must emphasize conscious experience. Sensory stimulation often affects one's mood without being detected or reaching conscious awareness. Although this might happen as readily with a melody as with an odor, olfaction may not necessarily be studied best as if it were analogous to hearing. People who have lost their sense of smell provide a good clinical example of this phenomenon. It is not until the sense is lost that one may become cognizant of all the subtle and subconscious effects of odor in everyday life.

In lecturing about the "pleasures of sensation" some years ago, Pfaffmann[251] called attention to a potentially important difference between sense modalities. Audition and vision tend to be keen senses as, for example, measured by the Weber fraction,[93] which indicates the physical amount by which a stimulus must be changed for a human to perceive the change. Variations in stimuli seem to have less effect on pleasure and displeasure as perceived by these senses than they would with taste and smell. For example, while the Weber fraction for vision may be about 10% for successively presented stimuli, for smell it is more than 25%. By contrast, the amount of pleasure or displeasure experienced through simple or unpatterned smell and taste sensations seems to be much greater than it is for vision and for hearing. For example, a color could hardly look as ugly as the malodor of hydrogen sulfide is repulsive. This chapter begins with a discussion of the keenness of the sense of smell and basic experimental psychology. This is necessary for an understanding of the more thorough discussion of the aesthetic aspects of odor perception.

PSYCHOPHYSICAL FACTORS

Weak Odors—Detection

The most commonly used psychological index of the effect of an odorous agent has been the threshold, which refers either to the boundary between detectable and undetectable concentrations (absolute threshold or limen) or to the differences between concentrations that can be detected and those differences that are too small to have any perceptual effect (difference threshold). Because of its apparent simplicity, the concept of threshold is widely used. The test subject is merely required to judge whether or not he is experiencing an odor or a difference in odor intensity in response to variations in concentration. This classic concept is associated with the beginning of modern experimental psychology, having been developed by Fechner in 1860.[109] It has also been a favorite of sensory psychologists ever since it was assumed that the subject's performance could be explained in terms of minimal neurologic response, i.e., a number of neural units firing to produce a conscious awareness of an external stimulus.

Although this concept is practical, there are substantial differences in individual responses, especially in odor thresholds. According to a review by the National Air Pollution Control Administration,[219] the odor threshold for hydrogen sulfide ranges from 1 to 45 ng/m^3 in air for individuals with different ages, sex, smoking histories, and places of residence. To date, most of the effort in this area has been devoted to determining the best method for measuring threshold and to selecting of judges for sensory panels.

Recently, the signal-detection theory has been judged as the most fruitful approach.[93] There are two salient differences between this theory and the classic notion of threshold. First, the contemporary approach questions the validity of the assumption that there is a certain cut-off point on the physical concentration scale above which there is and below which there is no conscious experience of odor.[93] Instead, signal-detection theory assumes that any concentration may be associated with conscious experience and that the subject's criteria for what constitutes an odor interact with odor intensity to determine his judgment in a specific trial. Second, classical threshold theory assumes that training of subjects or balancing of their tasks can eliminate human error or biases from the results. In contrast, signal-detection theory assumes that such bias is inherent in this form of decision process and must be measured in each situation. Usually, it can be measured by noting the proportion of false alarms, that is, incorrect affirmative responses ("I smell it"), to blanks. Although the dispute concerning whether or not detection is a continuous or discrete function of concentration is still debatable, there is convincing evidence that the likelihood of false alarm varies among observers, situations, and sense modalities in a manner predictable from knowledge about motivational aspects of the situation. Undoubtedly, the variation observed in odor thresholds is largely due to such factors.

In Sweden, sensory-chemical and meteorologic analyses are used to predict how often the concentration of an emission from a certain source, e.g., hydrogen sulfide from a pulp mill, may exceed the threshold for the odorant as it is dispersed in the surrounding area. However, these predictions sometimes underestimate greatly the evidence of odor as reported by local residents who have been instructed to make observations from their residences.[195] Although this may be partly explained by weaknesses in the dispersion calculations, evidence indicates that in such a situation a person is likely to overestimate both the incidence and duration of odor (see the discussion under "Socioepidemiology" on p. 104). Signal-detection theory indicates how one should correct the proportion of correct affirmative judgments, or "hits," of the presence of odor, with the proportion of false alarms. Since false alarms do not necessarily occur randomly, but as part of the expectation, motivation, and strategy of the test subject, it is not sufficient simply to subtract them from the proportion of hits. Correction must be made according to a model that relates these two variables to each other and to concentration. An index called d' is most often used. Although it first appears to be complex, the index is practical and straightforward. Also, it simplifies the selection of both methods, and judges and facilitates comparison of results from different experiments. The main practical consideration is the specification of a measure of response bias.

Signal-detection methodology has been applied to several experiments in the laboratory and in the field. Results suggest that sensitivity to a particular odor may be greater than those obtained with a classic method and also that individual differences are smaller. For example, Jones[154] obtained average threshold concentrations (i.e., concentration detected by the subject in 50% of the trials) of 4.5×10^{-5} and 1.69×10^{-7} in terms of molar ratio of *n*-propyl and *n*-butyl alcohol, respectively. By comparison, Corbit and Engen,[75] employing signal detection, obtained comparable values, ranging from 1.72×10^{-5} to 1.90×10^{-5} and from 0.39×10^{-5} to 0.60×10^{-5}, for the same odorants.

In an experiment with hydrogen sulfide, using the same basic detection procedure, Berglund *et al.*[29] found that a concentration of 7.37×10^{-7} mg/liter produced a proportion of hits of about 50% to 75% for both of the two observers. The apparent similarity of performance was much better than that for false alarm, which was generally $>30\%$ for one and $<15\%$ for the other. (For other detection values for various sulfur and nitrogen compounds in the laboratory and in effluents, see References 63, 125, and 193.)

The bias of human observers is considered indispensible to the production of results that are superior to more "objective" means of detection, such as the "artificial nose," both in general applicability and in sensitivity. It is also difficult to surpass the ability of humans to document rapid changes in odor. In an experiment involving odorous effluents from a mineral wool plant in a field study and dimethyl monosulfide in the laboratory, a signal-detection approach was used to advantage.[32, 33] The results showed that the ability of the subjects (housewives without any previous experience in odor research) to detect the odor reached a maximum only seconds after the odorant was presented in the test chamber. According to Lindvall (personal communication, 1975) no chemical method was sensitive enough in that situation.

Another study[194] was concerned with detectability of traffic odors over the several hours-long rush hours of a large city. Odor samples were collected on location and presented to untrained subjects via an olfactometer in a mobile laboratory. Despite the inevitable false alarms, the subjects produced a reliable and valid index, according to physical and chemical analyses of the air samples, between a busy city street and a relatively pollution-free university campus. Still another example of field application is presented by Grennfelt and Lindvall,[124] who monitored odorous effluents from a pulp mill at varying distances from the source.

Suprathreshold Odors—Psychophysical Scaling

The measurement of odor detection has in recent years reached a rather sophisticated level, both practically and theoretically. However, it is fair to assume that a weak and barely detectable odor is of relatively minor psychologic significance. At that level, the odor is not likely to have its characteristic quality, noted at moderate to strong levels, which could be associated either with pleasure and acceptance or with annoyance, rejection, or other negative effects. In general, sensory control of human behavior is not well represented by thresholds; in fact, it represents the level at which stimulus control of the psychological response breaks down. This is one of the important reasons why psychologists have devoted a great deal of attention and research to psychologic scaling. Another reason is that knowledge of how sensory magnitude increases as a function of physical magnitude (concentration) would contribute to the understanding of the operation of the transducers, in this case, olfactory receptors.

A great deal of scaling research done during the last two decades was stimulated by Stevens'[300] proposed psychophysical power law: perceived or psychological magnitude grows as a power function of physical magnitude. Earlier researchers had assumed either that Fechner's logarithmic law applied, or, worse, that one could scale perceived magnitudes as multiples of threshold concentration.[93] For olfaction, there has been little, if any, empirical support for either of these psychophysical scales. The assumption that equal increments of a physical unit, usually the threshold concentration, correspond to equal subjective increments has questionable validity. It does not hold for any specific compound, and, to make matters worse, may indicate different psychophysical relationships for different odorants. Therefore, the use of such a scaling procedure to compare odors is invalid. In their study of the intensity of odors from various effluents in a pulp mill, Berglund *et al.*[31] found that odors associated with the lime kiln and washery, which were measured in terms of concentration-dilution steps, were perceived as being much weaker than odors from the main stack, solving tank, and oxidation scrubber for all dilution steps. It was not possible to predict perception of odor intensity from such information.

In agreement with research in all the other senses, psychophysical scaling of odors has led to the conclusion that perceived intensity grows as a power function of odor concentration. The following mathematical formula was used:

$$\Psi = c^{\phi n}$$

where Ψ represents psychologic intensity, ϕ physical intensity (concentration), c the (arbitrary) choice of units of measurement, and n the exponent of the function. In over a dozen experiments on olfaction, this exponent, or steepness of the function when both physical and psychologic values are plotted in logarithmic units, was <1, varying between 0.07 and 0.7. This variation depended not only on the chemical compound, but also on the psychophysical method used and the differences among individuals.[28] The function $\Psi = 631^{\phi 0.45}$ was obtained when subjects compared subjectively the intensities of various concentrations of hydrogen sulfide (expressed as mg/liter at 20 C) with 2.29 mg/liter of acetone in air. All functions obtained by methyl mercaptan, dimethyl disulfide, and dimethyl monosulfide were also power functions, but with exponents of 0.18, 0.20, and 0.14, respectively. The same function seems to apply to complex odors from effluents. Thus, knowledge of the psychophysical function will undoubtedly contribute to the measurement of abatement of odorous air pollution.[31] The fact that the exponent is <1 means that the rate of increase in perceived intensity does not follow that of physical intensity. For example, when odor concentration is halved, odor intensity will be reduced much less, by an amount indicated fairly precisely by the exponent.

In the development of these scales,[93] subjects are generally asked to match odor concentration against numerals or other convenient quantitative dimensions, such as finger span. The procedure is similar to rating scales that use numbers of adjectives. The salient difference is that the scale is continuous and open on both ends so that the subjects are free to select any number or other magnitude in describing their judgment. Numbers, which are used most often, are matched to the odorant concentrations so that they are proportional to perceived or subjective magnitude. The central tendency of these numbers for a group of subjects constitutes the odor intensity scale. Its mathematical relation to the physical scale of concentration defines the psychophysical function, which usually fits the formula above.

The psychophysical scaling methods have been carefully tested in both the laboratory and the field. Because the psychophysical function obtained for butanol has shown such stability in results from different laboratories and from different techniques, it has been proposed for use as a standard reference scale.[226] Laffort and Dravnieks[180] have discussed the physical and chemical factors that may determine the size of the exponent. Other investigators have demonstrated the practical value of the approach in the field. For example, Svensson and Lindvall[303] have demonstrated the ability of human subjects to make so-called intramodal matches in olfaction, that is, to match the intensity of one odor against another. This ability has been used to evaluate odor intensity as a function of distance from stack effluents[124] and to assess the effectiveness of different spreading techniques in the reduction of manure odors.[196] Instead of numbers, the odors in some of these examples were matched by concentrations of hydrogen sulfide, which were controlled by the subject with an olfactometer.[95]

"Olfactory Fatigue"—Adaptation

The psychophysical function responds predictably to adaptation, masking, and mixing of odors. For example, the primary effect of adaptation is to increase the exponent n; that is, when a person's sensitivity is decreased through exposure to a constant odor, his sensitivity to various concentrations of this odor will increase faster with increases in concentration. In addition, there will be a decrease in the constant c, so that lower numbers or other matching magnitudes are now assigned to a lower concentration than they would be if the subject were not in an adapted state. The fact that adaptation affects the steepness of the psychophysical function means that the effect of adaptation is inversely proportional to concentration so that the weaker the odor the more it is affected.

Some investigators have reported[29] that olfaction is so greatly influenced by adaptation that exposure to an odor for a matter of minutes will cause the odor to disappear through "olfactory fatigue." It is not that simple or dramatic, fortunately. A good rule of thumb is that the most effective variable of adaptation is concentration or strength of the odor. This sense modality can be regarded as a signal noise system, where sensitivity to odor (signal) is determined by the adapting effect of the odor present (noise). Any change in the odor quickly (probably in a matter of seconds) decreases the sensitivity of the system so that a stronger odor or signal is now required for a person to detect it. However, the duration of exposure to the odor does not seem to affect sensitivity as greatly. Those who have assumed that constant exposure causes odor sensitivity to become nil have probably failed to distinguish adaptation from habituation, which refers to a diminution of response and attention to a stimulus whose consequences seem unimportant to the observer.[98]

In a very pertinent experiment, Ekman *et al.*[92] asked observers to estimate the perceived intensity of 0.7, 0.9, 2.6, or 4.6 ppm of hydrogen sulfide. Under constant exposure to these concentrations, the perceived intensity initially tended to decrease rapidly and exponentially, but then reached an asymptotic level where the odor remained. Only one subject indicated that the odor disappeared: this was the only finding that occurred as predicted from the belief that olfaction shows complete adaptation.

The substances that stimulate the olfactory system (mediated by the first cranial nerve) also may stimulate the trigeminal nerve (mediated by the fifth cranial nerve),[312] which is assumed to convey sensory information about pain and irritation, e.g., from exposures to ammonia. A person may find it difficult to distinguish whether one or both responses are reacting. Some pollutants, such as aldehydes, involve both.

Cain[58] has shown that unilateral destruction of the fifth nerve contributes significantly to the intensity of an odor. He also observed that the trigeminal nerve seems to be less affected by adaptation than by olfaction.[57] The implication is clear: one cannot depend on adaptation to reduce awareness of odorous pollution.

Odor Mixture—Perceived Intensity

The adaptation situation can also refer to self-adaptation, the effect of an odorant on the ability of an observer to detect or discriminate its odor. Cross-adaptation refers to the effect of odorant A on an observer's sensitivity to the odor of B. In general, it leads to the same kinds of effects as self-adaptation, with the following qualifications. First, the effect is not necessarily transitive or symmetric. Odorant A, used as an adapting odorant, may affect the perception of B more than B affects A.[56] This undoubtedly relates to the nature of the effect of the odorant on the olfactory receptors, which is as yet only poorly understood.

Another puzzle is that of facilitation, which is observed occasionally in some subjects. Exposure to one odorant immediately before exposure to another increases the perceived intensity of the second odorant.[75, 97]

In some tests, odors are presented simultaneously (assuming chemically inert mixtures), while in the adaptation cases they are presented successively. Mixtures, for example, A and B in liquid solution, smell stronger than either A or B alone. However, the perceived intensity of the mixture is always less than the sum of those of the individual components as measured on a psychophysical magnitude scale.[155] Instead, the psychological result is a weighted average of the two.[34] Although space limitations preclude the inclusion of descriptions of the mathematical model and psychophysical theory involved, the importance of their practical results should be stressed. Simply stated, one odor tends to dilute or mask the other odor in a mixture of two in a predictably quantitative manner. This principle appears to be general, in that it probably applies to more than two components. Thus, the more components, the weaker the mixture. When mixing two components, the psychologically qualitative differences between the components tend to yield a weaker odor than when two similar odors are mixed, e.g., pyridine and linalyl acetate versus pyridine and hydrogen sulfide.

These results suggest that a deodorizer should be a complex odorant containing many qualitatively different odorants that mask any malodor. However, these principles have not as yet been studied thoroughly with complex mixtures. Research[58, 59] has mainly involved two odorants at a time, one a so-called malodor and the other a masking odor in efforts to reduce industrial odors.[161, 224] The principles of odor mixing that have been developed in the psychology laboratory apply to air pollution. They are perceptual and do not indicate any obvious physical or chemical interaction between the odorants. Indeed, neither the physics nor the chemistry of the interaction is at all understood.

There is also the possibility that mixtures of odorants may produce synergistic effects that may not necessarily be detected by the trigeminal nerve or olfactory response. For example, Modica *et al.*[222] reported that mine workers who smoke have been adversely affected by their environment. There are similar problems involving the effects of alcohol and carbon monoxide. Increasing the number of components of an air-polluting source probably does not increase its odor intensity; it may in fact decrease it, while at the same time producing other effects.[27]

Deodorizing—Perceived Quality

One anecdote describes olfaction as analytic, i.e., by using it, an individual can sort out the components of a mixture. This involves the quality of the odor rather than its intensity. However, the description of this modality as analytic rather than synthetic is not easily verified. This kind of assessment is difficult because the mixing of odors produces more than one psychological effect.[28]

When a deodorizer like lavandin oil is added to a malodor like pyridine, counteraction commonly results, i.e., the quality of the malodor is reduced.[59] There are differences in the capacities of particular odors to counteract each other.

A high concentration of a deodorizer may completely mask a malodor, but it may be so strong as to be unpleasant itself. In counteraction, the overall odor intensity is reduced; thus, the combined effect of the two odors is less than the arithmetic sum of the two. The classic literature stressed the debate about whether or not there is compensation, as when the sum of the perceived odors of a mixture is less than any of the individual components, or when odor may be eliminated altogether. Although compensation has been reported for certain combinations,[177] it is far from commonplace.

Another observation is that the odor of the mixture tends to dominate the odor of the weak malodors, i.e., the mixture is stronger but also more pleasant. When a malodor is very strong, addition of the deodorizer at the same concentration as in the previous example produces a decrease in the intensity of the malodor, thereby increasing the pleasantness. At this level there is counteraction. If now the deodorizer is kept constant and the malodor is increased, the malodor appears to grow faster than the intensity of the mixture of the two. The important implication is that the effectiveness of the deodorizer will be inversely proportional to the intensity of the malodor. Cain and Drexler[59] point to two practical problems associated with the application of the data on this dose-response relationship. One is that deodorizers will not be effective against strong malodors; the other is that the concentration of the deodorizer needed in a particular situation cannot be predicted. Counteraction or masking should not be regarded as general cures for bad odor, because either means adding more substance to the air with generally unknown effects.

AESTHETIC AND HEDONIC FACTORS

Coding Feelings and Emotions

It is commonly agreed that the most important effects of odors are perceptual, especially the effects of pleasure and displeasure. In addition to the restriction of the term to motivational and emotional effects rather than criticism or appreciation of abstract beauty, the aesthetics of odor tend to be dominated by displeasure. Although perfumes, food, and flowers have pleasing odors, only 20% of the estimated 400,000 odorous compounds are pleasant.[130] Part of the reason for this uneven split may be that people seem to have a strong tendency to judge any unfamiliar odor as unpleasant. In one study subjects were presented with 110 diverse odors[100] about half of which were unfamiliar to them. Of these unfamiliar odors, only 11% were judged as pleasant, 50% as unpleasant, and 39% as neutral. These are average percentages for individuals. They do not indicate that individuals agreed on which odors were pleasant and which were unpleasant. Also, an odor that was familiar to one was not necessarily familiar to another. Perhaps suspicion of the unfamiliar is what characterizes the sense of smell, alerting and warning people and putting them in a state of arousal rather than relaxation. Of course, not all familiar odors are pleasant. Gloor[120] has suggested that the sense of smell may through evolution have played a role in helping to get animals beyond the simple reflexive behavior mediated at the hypothalamic level.

There are rather convincing data showing that human preference for odors is largely absent at birth. The number of hedonic responses to odors increases with age.[93] A particular child may show likes and dislikes of certain odors, but the reactions of children cannot be predicted from the preferences of adults. In general, children seem to be more tolerant of odors than are adults.

Pleasant and Unpleasant Odors

Cultural as well as age differences affect preferences for odors such as that of perfumes. In the Western culture, adults complain about the odor of cattle manure.[196] For reasons that are not obvious, many people enjoy the smell of a barn but not of the outdoor privy. No amount of familiarity seems to change this. In Sweden, psychological scaling methods used to measure the degree of annoyance with such odors resulted in legislation limiting the odorous emissions from the combustion toilet.[198]

Studies of cellulose factories have revealed that among the main offensive odors resulting from industrial processes are hydrogen sulfide, methyl mercaptan, dimethyl sulfide, and dimethyl disulfide. There are also unpleasant odors associated with combustion engines (especially diesel), refineries, synthetic resin reactors, printing enamel plants, food processing plants, soap factories, and garbage dumps.[171, 307] According to a report from Monsanto,[277] among the most unpleasant odors associated with people and their homes are proton acceptors or donors, including carboxylic acids in sweat and rancid foods, thiols, phenols, amines, and tobacco smoke.

In Britain, Moncrieff[223] had various groups of people rate a diverse sample of both pleasant and unpleasant odors. Among adults, natural odors such as fruits, vegetables, flowers, and spices were preferred over others. The most disliked odors were again such odors as pyridine, butyric acid, phenol, and ethyl mercaptan. There is apparently wide agreement in these judgments, barring of course individual differences associated with unique experiences. In general, the odors of natural materials are preferred to those of synthetics, for the reason, Moncrieff assumes, that their molecules are complex. (Of course, natural materials contain many types of molecules, which may be simple when taken individually.) Moncrieff also points out that the higher the concentration, the less pleasant the odor.

The researchers at Monsanto claim to have discovered a "fresh air" smell that counteracts malodors without affecting pleasant odors. Unfortunately, the company is filing for patent production and will not divulge the secret behind this astonishing effect. The method that Monsanto has suggested, however, has been severely criticized.[244]

Do Odors Affect Health?

There have been a number of socioepidemiologic studies of the extent to which people are aware of and bothered by odors. Complaints about odors range from 27% in rural areas to 78% in urban areas.[191] Even at distances greater than 20 km from the source of the odor, about 30% of the respondents in a survey in Sweden and the United States had such complaints.[64, 113, 114] Commenting on the American results, Shigeta[278] points out that although very few industrial enterprises were actually violating pollution regulations, over 30% of the complaints received were still about malodors. This may once have led one to dismiss the complaints as subjective and useless. However, there has been a change in the official attitude toward such information. In Sweden,[192] great emphasis is placed on reports of annoyance related to environmental factors. Such reports are used as a basis for intervention by authorities. Specifications of physical or chemical concentrations and composition may be less reliable indicators than the human nose. It is believed by many that odor survey techniques with untrained observers can provide reliable results.[141] Mobile laboratory facilities, which are used in several countries, enable investigators to expose subjects to measured quantities of air in any locality.[193, 294]

There has also been a change in attitudes about the environmental norms for health. It is often argued that an increased standard of living should be free of *all* disturbance from environmental factors—not just those involved in causing disease. A purely objective medical approach is being questioned and support given the World Health Organization's definition of health as "A state of complete physical, mental, and social well-being and not merely the absence of disease or infirmity."[315]

There is, however, no evidence that the experience of malodor *per se* produces disease. Epidemiologic studies are needed to document the effect of purely psychological factors. This argument is not pursued in this document, but it should be kept in mind that poor health may in turn increase the displeasure or at least the frequency of complaints about odor.[96]

Socioepidemiology—The Measurement of Attitude about Odor

In socioepidemiologic methods, statistically defined populations are queried regarding odor perception and its effect on well-being. According to the conclusion drawn in the Fourth Karolinska Symposium,

> The classes of variables which are most relevant to annoyance surveys include: Level of awareness of sources of environmental pollutants, feelings or affective responses to these sources, duration or periodicity of the reaction, salience of the response, and demographic, sociological, and economic characteristics. Other related variables are information level and feelings about environmental problems in general, social awareness of annoyance issues, and attitudes toward the source of pollution, such as beliefs about its potential for harmful effects.[197]

Despite the need for human test subjects, the problem of subjectivity does remain. For example, in a survey of public opinion about diesel exhaust odor, Springer and Hare[294] observed that "highly concerned citizens might rate the odors they perceive as being more objectionable than they actually are in the hope that they can thereby strike a blow at pollution in general." In this case, as in the detection procedures described above, one must include a measure of such bias. Petitioners in a public health case reported that they were annoyed up to 50% more than the "silent" majority of the population.[64] On the other hand, those making their living at a factory causing odor pollution are less likely to report annoyance. Although this could be explained in terms of adaptation or habituation, response bias is probably also a factor.

Another striking example of how motivation or attitudes may affect epidemiologic data is described by Cederlöf *et al.*,[65] who studied noise associated with airports. These investigators knew from earlier work that people in a certain city in Sweden were bothered by noise from commercial and military planes. They divided a group of 270 people in two halves. One half, the experimental group, was provided with interesting information about flying and airplanes, including a book presenting a favorable 50-year history of Swedish military aviation. The other half of the respondents, the control group, were given no such special attention. When the survey began, about a month later, 43% of the control group reported being very disturbed by the airplane noise, compared with 18% in the experimental group. A commonly used method of relying on spontaneous complaints, such as letters to the editor and the like, is risky. It is likely to reflect bias plus all the other problems incurred with unrepresentative sampling.

Individual differences are inevitable problems in any method that relies on human test subjects. There were even sex-, health-, and age-related differences in annoyance to odor reported by a group living within 3.2 km of a sulfate pulp mill.[113, 114] In another study, which was concerned with the relationship of personality traits and attitudes to the adverse effects of odors, a form was used to measure propensity to neurosis. Results indicated that there was indeed a correlation between these two traits.[64] In addition, it was found that annoyance with odor was occasionally combined with reports of nausea and headaches. The authors concluded that

> The results also show that the annoyance is due not only to the exposure in question but also to factors among those exposed. Thus it is clear that annoyance is more frequent among those reporting previous respiratory or cardiovascular diseases and also among persons with a propensity to neurosis, sensitivity to other external environmental factors and propensity to displeasure with other aspects of the community.[64]

When analyzing socioepidemiologic data, it is tempting to interpret the frequency of response as indicating intensity of annoyance. One must distinguish between the existence of an effect and its magnitude. The same percentage of people in two different cases may report displeasure with malodor—one response having been elicited by a strong odor, the other by a weak odor. Methods aimed primarily at opinion and attitude should not be relied upon to measure perceptual magnitude. Currently, there is research aimed at bridging this gap by adapting methods used in psychological scaling to the problems of the epidemiologist.[30]

Psychological Scaling—The Measurement of Perception

To measure the magnitude of aesthetic effects, the scaling methods used to discriminate suprathreshold odors can be used. They are flexible and well understood because of the great amount of laboratory research that has been done. Although they were originally developed for psychophysical scaling of intensity when there is a physical or chemical basis of the sensation, these scaling methods have been extended to situations where there is no precise knowledge of such correlations, as in the case of odor pleasantness.

In general, the pleasant or unpleasant aspects of an odor seem to be more important than its intensity. In one experiment, subjects were asked to judge the pleasantness of a diverse sample of odors by assigning numbers to them proportional to pleasantness; that is, the more pleasant, the higher the number.[99] While the dynamic range, or range from weakest to strongest concentration for odor intensity, is about 10:1 for a typical odorant, the range obtained in this case for pleasantness was as much as 150:1. Of course, both figures depend on the odorant samples; however, they do provide a rough indication of the difference. Many investigators have concluded from various experiments that the variability in pleasantness is the outstanding characteristic of odors.[274, 337, 339]

It is, in fact, difficult to ignore the hedonic attribute of an odor when the task is to judge its intensity. Referring back to the psychophysical power function, it seems that the unpleasantness of odor increases the intercept (c) and decreases the exponent (n, or steepness in log-log coordinates); i.e., when an odor is unpleasant it tends to be judged strong at all concentrations.[94] It is not possible to describe precisely the relationship between odor intensity and concentration; but, as previously noted, pleasantness seems to decrease as intensity increases.[135, 227] At very low concentrations, when the odor is hardly detectable, it tends to be judged neutral. As the concentration of pleasant odors (e.g., fruit or food) is increased, pleasantness increases at first. With still further increases in concentration, pleasantness reaches a maximum, a plateau, and might even decrease. Odors that tend to be unpleasant at any concentration, such as hydrogen sulfide, will be judged more and more unpleasant as concentration is increased.

Odor Pleasure and Physiologic State

Cabanac[55] has shown that the physiologic state of a person, as determined, for example, by hunger versus satiety, will determine judgments of pleasantness, whereas judgments of intensity, which depend on external stimulus factors, are independent of this state. From recent research in the Brown University laboratories comes the important qualification that physiologic state mainly affects food-related odors.[228] Although odor of a food may be pleasant before eating, but unpleasant after one has overindulged, a person's judgment of the intensity of that odor would remain the same. On the other hand, when the person's physiologic state is unchanged, judgment of pleasantness may be relatively stable compared with intensity judgments.[267]

As a rough rule, changes in the internal environment will affect primarily a person's hedonic reaction to odors; the effect of external factors, such as pollution, will affect the ability to detect and evaluate the strengths of odors. Although there are no rigidly defined categories of internal and external effects of stimulation, the difference must be considered.

The Effect of Stimulus Context

In situations without special physiologic effects or emotional consequences, an odor that is unpleasant at first will appear less unpleasant after some exposure or familiarization. The opposite may happen with a pleasant odor. It is as though the hedonic value of odors regresses toward the neutral zone of the scale. The same seems to happen to the position of a particular odorant in a set that has been singled out for special judgment (W. S. Cain, 1975, personal communication). For example, hydrogen sulfide does not seem as unpleasant after one has been exposed to it in a detection or discrimination experiment.

Measurement of Displeasure

In one method to measure displeasure with an odor, a target odor is matched with a concentration of an odor, such as that of hydrogen sulfide or pyridine, that is found unpleasant regardless of concentration. There are two parts to this method. First, a psychophysical scale for hydrogen sulfide is developed by using a scaling procedure similar to that described above for suprathreshold odors. Next, the test subject matches the target odor to a concentration of hydrogen sulfide by manipulating the concentration in a dilution system. When odor intensity rather than pleasantness is being tested, the standard scale mentioned above should be used.[226]

In one study, air from a fertilized field was compared with a range of hydrogen sulfide concentrations. Investigators used a magnitude estimation technique in which groups of data were transferred to a common scale.[196] A similar intramodal matching technique has also been used to evaluate the kitchen odors of cooking cabbage and onion when matched to pyridine diluted in water.[195] In this way, odor abatement in the field can be evaluated. The best method is the one that produces the lowest concentration match of hydrogen sulfide or pyridine—thus, presumably, the least displeasure. These methods will play a larger and larger role in evaluating the aesthetic aspects of the environment and will undoubtedly result in more precise predictions of perceptual magnitude.

9

Summary and Conclusions

OCCURRENCE, PROPERTIES, AND USES

Hydrogen sulfide is widely distributed among a variety of man-made and natural settings where sulfur-containing organic matter may decompose anaerobically. Examples of such settings are sewers, sulfur springs, volcanic gases, and deposits of coal, petroleum, and natural gas. The sulfur content of natural gas, however, varies widely—even in the United States. It ranges from almost none up to 40%. Hydrogen sulfide may be increasingly recognized as a hazard that results from tapping geothermal energy sources. The gas is also generated as a by-product of or waste material from the process of removing sulfur from fossil fuels and from the production of carbon disulfide, coke, manufactured gas, thiophene, viscose rayon, and kraft paper. Eventually these "off" gases are converted to elemental sulfur, the form most convenient for storage and handling, or to sulfuric acid, if a local market exists. Large quantities of hydrogen sulfide are used in the production of heavy water for atomic reactors, and sodium sulfide is used widely in the preparation of hides for tanning.

Hydrogen sulfide is a liquid at temperatures above −83 C and a gas at temperatures above −60 C. It is both flammable and explosive at concentrations from 4% to 46% in air. It is heavier than air, and is soluble in both polar and nonpolar solvents. Aqueous solutions are unstable unless oxygen is rigidly excluded. Hydrogen sulfide has two acid dissociation constants:

$$H_2S \rightleftharpoons HS^- + H^+ \quad pK_a = 7$$
$$HS^- \rightleftharpoons S^= + H^+ \quad pK_a = 12$$

Thus, at the physiologic pH of 7.4, about a third of the total sulfide exists as the undissociated acid (H_2S) and about two-thirds as the hydrosulfide anion (HS^-). Only infinitesimal amounts exists as $S^=$. Even very alkaline solutions of sodium hydrosulfide or sodium sulfide tend to distill off the hydrogen sulfide slowly.

THE SULFUR CYCLE

Microorganisms are ultimately responsible for the biogenic hydrogen sulfide in the atmosphere, but bacteria can both reduce and oxidize various components of the sulfur cycle. In the upper atmosphere sulfide is oxidized to various sulfur oxides, but sulfate is recognized as the main form in which sulfur is transported in geochemical cycles. A substantial portion of the lower atmospheric sulfate, however, is derived from ocean sprays.

Sulfate deposited on the earth may be reduced to the equivalent of hydrogen sulfide by plants and incorporated into their proteins. Herbivorous animals transform plant protein to animal protein. The eventual decay of both types of proteins, as mediated by microorganisms, results in the evolution of hydrogen sulfide.

The combustion of fossil fuels in intensive industrial activities in the northern hemisphere generates 37% of the total atmospheric sulfur. It is estimated that by the year 2000 the anthropogenic and biogenic sources of sulfur will become equal, but what impact that will have on the world sulfur cycle is unknown.

FATE OF HYDROGEN SULFIDE IN ANIMALS AND HUMANS

Although the principal salt of commerce, sodium sulfide nanohydrate ($Na_2S \cdot 9H_2O$), probably has the corrosive potential of lye, it is unlikely that it is absorbed through the intact skin. Hydrogen sulfide is absorbed through the skin, but only during very intense exposures. Intoxication in humans invariably results from inhalation of hydrogen sulfide gas. When soluble sulfide salts or solutions of hydrogen sulfide are given by routes other than inhalation to laboratory animals, the gas is easily detected in the expired breath. Pulmonary excretion may be a quantitatively important means of terminating the systemic toxic effects of hydrogen sulfide, but additional studies are needed to resolve conflicting findings in the literature.

There is evidence that both enzymatic and nonenzymatic oxidative biotransformation pathways exist in mammalian species. Sulfide that is not excreted via the lung is probably converted to thiosulfate or sulfate in the body. Some investigators believe that hydrogen sulfide is constantly generated in the human gastrointestinal tract, then rapidly absorbed and metabolically inactivated. The literature contains conflicting reports about the presence of hydrogen sulfide in normal human flatus. The suggestion that hydrogen sulfide may accumulate under various pathophysiologic conditions, such as intestinal obstruction, deserves further study. Both hydrogen sulfide and methyl mercaptan, which have been detected in the parts per billion range in normal human breath, are associated with oral malodor. Methyl mercaptan may be responsible for fetor hepaticus, the unpleasant odor found in the breath of patients with severe liver disease.

EFFECTS OF HYDROGEN SULFIDE ON ANIMALS

Experimentation on the biologic effects of hydrogen sulfide has been spread thinly over two centuries (see Appendix II). The identical stimulatory effects of sulfide and cyanide on respiration are now understood in terms of their activation of carotid body chemoreceptors. The resulting hyperpnea is eventually replaced by respiratory depression and apnea. The latter effects are mediated through the brain stem nuclei. Hydrogen sulfide is about as acutely toxic as hydrogen cyanide.

The observed effects of sulfide on the blood *in vitro* have created considerable confusion in the scientific literature. An impairment of the oxygen transport capability of the blood plays no role in acute sulfide poisoning. Rapid generation of sulfhemoglobin *in vitro* involves exposing the blood to concentrations of hydrogen sulfide that are incompatible with life because of their effects on respirations. Sulfhemoglobin, as generated *in vitro*, does not appear to be related to the "pseudosulfhemoglobin" generated *in vitro* or *in vivo* when blood is exposed to "oxidant" drugs and chemicals.

The key lesion in acute sulfide poisoning, as in acute cyanide poisoning, is an inhibition of cytochrome *c* oxidase. Both sulfide and cyanide form stable but dissociable complexes with the ferric heme iron of methemoglobin. The properties of the sulfmethemoglobin complex are distinctly different from those of sulfhemoglobin, but at least one similarity is known—both pigments are unstable and tend to decompose to hemoglobin. The induction of methemoglobinemia affords significant protective and antidotal effects in the acute sulfide poisoning of animals. The procedure has been used successfully in one severe human intoxication.

In cases of acute sulfide poisoning, artificial respiration may be of value in accelerating the pulmonary excretion of hydrogen sulfide. Oxygen has an important therapeutic role in the subacute syndrome where pulmonary edema is apt to supervene, but it is not a specific antidote. Almost nothing is known about the chronic effects of sulfide in animals, particularly in relation to sulfhemoglobin formation.

EFFECTS OF HYDROGEN SULFIDE ON HUMANS

Hydrogen sulfide intoxication has been classified under three rubrics: acute, subacute, and chronic. Acute intoxication is a dramatic, systemic reaction resulting from a single massive exposure to $>1{,}400\ \mu g/liter$ (1,000 ppm) of hydrogen sulfide in air. This condition is characterized by rapid (often instantaneous) loss of consciousness followed by convulsions and respiratory failure caused by the paralyzing effects of the gas on the centers of respiration. Death due to histotoxic anoxia is the frequent outcome of acute intoxication unless resuscitation is begun immediately.

Subacute hydrogen sulfide poisoning is a localized response to the irritant properties of the gas following continuous exposure to concentrations between 140 and 1,400 μg/liter (100 and 1,000 ppm). Eye irritation, manifested as conjunctivitis, keratitis, or both, is the most common form of subacute poisoning. Respiratory tract irritation is also an effect of subacute poisoning. If exposure is prolonged, irritation of the deeper regions of the lung may cause pulmonary edema. It is important to emphasize that, at these concentrations, hydrogen sulfide produces rapid paralysis of the olfactory apparatus, thereby neutralizing the sense of smell as a warning system.

There is no unanimity of opinion among authors as to whether chronic hydrogen sulfide poisoning represents a discrete clinical entity. Some believe that the signs and symptoms collectively referred to as chronic poisoning actually represent recurring acute or subacute toxic exposures.

In the management of the acute syndrome, the mechanical assistance to ventilation may have some advantages over positive-pressure oxygen in that pulmonary excretion may be an important route for the elimination of absorbed hydrogen sulfide. Oxygen, however, is specifically indicated if pulmonary edema supervenes. The therapeutic induction of methemoglobin has been employed with apparent success in at least one severe human poisoning.

EFFECTS OF HYDROGEN SULFIDE ON VEGETATION AND AQUATIC ANIMALS

Susceptibility to hydrogen sulfide appears to vary little among animal species. In contrast, plant species vary widely in their sensitivity to its toxic effects. Some plant species, e.g., lettuce and sugar beets, actually show growth stimulation at concentrations that result in damage to other plants [0.04 μg/liter (0.03 ppm)], but all plants show deleterious effects if the exposure is sufficiently intense [0.4 μg/liter (0.3 ppm)]. Sulfide taken up by plants is metabolized primarily to sulfate or incorporated into plant proteins. Experiments with algae suggest that different metabolic processes are responsible for differences in their susceptibility. Fish are more susceptible to sulfide in acidic environments perhaps because low pH favors the undissociated form (H_2S), which more readily penetrates the membranes of the fish. The biochemical basis for sulfide toxicity in plants is not understood, but inhibition of cytochrome *c* oxidase does not appear to have been ruled out.

AIR QUALITY STANDARDS

Air pollution by hydrogen sulfide is not a widespread urban problem. It is generally confined to the vicinity of emitters such as petroleum refineries, kraft paper mills, industrial waste disposal ponds, sewage treatment plants, heavy water plants, and coke ovens. The odor threshold for hydrogen sulfide lies between 1 and 45 mg/m^3 (0.7 and 3.0 ppm). At these concentrations no serious health effects are known to occur.

Standards for ambient air quality (as well as occupational exposures) have been set up by several states and foreign countries. At least eight countries have adopted emission standards. In the United States the threshold limit value has been set at 15 mg/m^3 (10 ppm) for a 8-hr workday and a 40-hr workweek. No national ambient air standards have been adopted for the United States.

THE PSYCHOLOGICAL AND AESTHETIC ASPECTS OF ODOR

Even though very little is known about the long-term health effects of exposures to hydrogen sulfide, there are obvious aesthetic aspects which are probably also relevant to other odiferous pollutants. In years past the measurement of odor thresholds has been the most common perceptual approach to making such studies quantitative. However, the agreement among various studies, including those with hydrogen sulfide, has been very poor. Scaling studies, in which the increase in odor intensity is related to concentration, have resulted in much better agreement. The results suggest that a power function describes this modality reasonably well as it also does for the senses of sight and hearing.

Adaptation to odor is a phenomenon distinct from habituation. It appears to be related primarily to the concentration of the odoriferous compound rather than to the duration of exposure. When several odors are present in a mixture simultaneously, they tend to dilute or mask each other. Thus, the perceived intensity of a mixture of odors is usually weaker than the intensity predicted from the arithmetic sum of the intensities of the components. There are exceptions to this general rule, however. Some odors may have synergistic intensities.

Odor preferences are not evident at birth, but are most likely to be learned—perhaps as a survival function. No evidence exists that malodor *per se* produces disease, but poor health may increase the displeasure with odor. Anxiety over the possible cause of an odor may produce severe discomfort. The pleasantness or unpleasantness of odors are probably not inherent characteristics of the compounds or stimuli, but are determined primarily by physiologic and psychological factors in the person perceiving them.

SAMPLING AND ANALYSIS

There are a variety of methods for analyzing hydrogen sulfide. Some are suitable for field studies, and others for the most sophisticated trace analyses. The staining of lead acetate paper strips is a technique long used in the field. The best spectrophotometric technique involves the reaction of sulfide with *N*,*N*-dimethyl-*p*-phenylenediamine and ferric chloride to form methylene blue. More recent approaches include a variety of gas chromatographic techniques and a silver-sulfide selective ion electrode. Improved analytic techniques, however, especially for continuous monitoring, are always desirable.

10

Recommendations

1. Exposure to high concentrations of hydrogen sulfide can create an extreme medical emergency. In such emergencies, mechanical assistance to aid respiration should be instituted immediately when indicated. Positive-pressure oxygen may be specifically required if pulmonary edema appears. Additional clinical experience with the therapeutic induction of methemoglobinemia is desirable.

2. In the English-language literature, there are no satisfactory studies on the long term (>30 days) effects of exposure to low concentrations of hydrogen sulfide. In future research, high priority should be given to this area. Particular emphasis should be placed on the possible accumulation of abnormal blood pigments in laboratory animals exposed to low concentrations.

3. Additional studies should be directed toward elucidation of the biologic fate of hydrogen sulfide in humans and laboratory animals. The importance of pulmonary excretion as compared to other mechanisms for inactivation should be determined. Precise biotransformation pathways should be defined for common laboratory animals with a view, again, toward more rational management of the acutely poisoned victim. Exposure conditions that trigger sophisticated biologic parameters, such as bronchoconstriction, need precise definition.

4. The role of hydrogen sulfide in the global sulfur cycle should be continually surveyed and evaluated. Human activities that result in anthropogenic hydrogen sulfide threaten to intrude in a major way on the sulfur cycle within the very near future. Also foreseen are additional stresses resulting from the current energy crisis.

5. Studies should be directed toward the effects of hydrogen sulfide on vegetation so that the physiologic and biochemical bases for those effects can be understood. Dose/response relationships should be established for plant damage to determine whether or not clear-cut thresholds exist.

6. Eventually, after more data have been accumulated, the establishment of national ambient air quality and emission standards for hydrogen sulfide should receive consideration.

7. Psychophysics and socioepidemiology should not be neglected as possible approaches to such questions as: What are the long term psychological implications of air pollution by odiferous compounds? Is

such pollution associated with prolonged or permanent adaptation of the sense of smell? If anosmia occurs, what is its time course, what agents are involved, and what are the chances for recovery?

8. Highly specific and sensitive analytic techniques for hydrogen sulfide in air and water should be developed, particularly those which lend themselves to continuous monitoring.

Appendix I
Hydrogen Sulfide—Sampling and Analysis

The perception threshold for the characteristic rotten egg odor of hydrogen sulfide varies considerably. Depending on individual sensitivity, it can range from <0.028 to ~0.14 μg/liter (<0.02 to ~0.10 ppm) at 25 C and 760 torr. Adams and Young[4] have reported odor detection thresholds of 0.01 to 0.045 μg/liter (9 to 45 $\mu g/m^3$). Consequently, the odor of this gas can be a very sensitive indicator of its presence in low concentrations. Only the most sensitive analytic methods can be used to determine the concentrations at the lower range of the odor detection threshold.

The sampling and analytic methods for hydrogen sulfide that are used in ambient air pollution studies and in industrial hygiene surveys are based on a variety of chemical and instrumental techniques. These include iodometric titration, chemical reaction and conversion to methylene blue or molybdenum blue, impregnation of paper tape or tile with lead acetate, reaction with a silver membrane filter, gas chromatographic methods, coulometric or galvanic methods, and methods using a selective ion electrode.

The iodometric method is based on the oxidation of hydrogen sulfide by absorption of the gas sample in an impinger containing a standardized solution of iodine and potassium iodide. However, this solution will also oxidize sulfur dioxide, which is usually present in the contaminated ambient air. Both gases are stable when mutually present in low concentrations. The unreacted or excess iodine is estimated subsequently by titration with standard sodium thiosulfate solution. Sulfur dioxide may be oxidized separately to sulfuric acid by a dilute acid solution of hydrogen peroxide. Hydrogen sulfide will not interfere if the solution is acid. Application of the iodometric method in industrial hygiene surveys has been described by Jacobs.[149]

Another variation of the iodometric method is to pass a known volume of air through a solution of ammoniacal cadmium chloride contained in two bubblers in series. The collected samples are stripped by aeration of any sulfur dioxide that may have been trapped. The cadmium sulfide precipitate is then dissolved in concentrated hydrochloric acid. This solution is titrated with standard iodine solution, using starch as an indicator. Cadmium acetate may also be used as the absorbing solution.[148] Iodometric methods are suitable mainly for industrial hygiene surveys. Their accuracy is only about 0.70 μg/liter (0.50 ppm) of hydrogen sulfide for a 30-liter air sample.[148]

Paper tape or tiles impregnated with lead acetate are the basis of a common method for the routine measurement of low concentrations of hydrogen sulfide in the atmosphere. The unglazed, impregnated tiles are exposed at selected locations and protected from rain. After exposure, the shade of the tiles is compared with known standards to estimate the concentration of hydrogen sulfide. This method gives only an indication of the relative exposures to hydrogen sulfide in various localities.[118, 148] The exposed, darkened tiles fade on exposure to air turbulence and light. Because the discoloration of these tiles will eventually fade, the period of exposure should not be greater than a day or two. The range of average concentration that can be determined by measurement of the surface absorption of the lead acetate is between ~0.150 and ~1.5 μg/liter.[67, 118]

In field studies of air pollution, continuous measurements of the hydrogen sulfide content of the atmosphere have been made by automatic samplers in which a measured air volume is filtered through lead acetate-impregnated filter paper tape. The optical density of the dark colored spots of known area is compared with a standard, unexposed, impregnated spot of similar area. Studies by Sanderson *et al.*[266] have shown that relatively large measurement errors can occur due to the fading of the dark, lead sulfide spots by the action of light, sulfur dioxide, ozone or oxidant, and by any other substance capable of oxidizing the lead sulfide surface. Because this fading can occur in a short time, a negative result may not be indicative of the absence of hydrogen sulfide in the air. However, High and Horstman[139] have reported results with this tape sampler that were in reasonably good agreement with the methylene blue method for hydrogen sulfide. The lead sulfide stains did not fade significantly when the tapes were stored in vapor- and moisture-proof bags during an 8-week period.

In an improved method proposed by Paré,[245] paper is impregnated with mercuric chloride instead of lead acetate. He has reported that the mercuric chloride paper tape is sensitive and reliable for the measurement of hydrogen sulfide in air and that the resultant spots are stable even in the presence of high concentrations of sulfur dioxide, oxides of nitrogen, and ozone. Sensitivity was adequate in the range of 0.700 μg/liter. However, Dubois and Monkman[89] confirmed that the spots on mercuric chloride tape are resistant to fading effects but found that the presence of sulfur dioxide in the air causes a substantial change in the hydrogen sulfide threshold of the tape.

A method for the determination of hydrogen sulfide, based on passing a measured volume of air through a silver membrane filter, has been studied by Falgout and Harding.[106] The resultant formation of silver sulfide causes a decrease in the reflectance of the silver surface that is proportional to the hydrogen sulfide exposure. This method is also sensitive to the presence of mercaptans in the air. A variation of this silver exposure method involves the use of silver coupons and subsequent removal of the sulfide after exposure, followed by chemical analysis by the methylene blue method.

Various detector tubes containing inert particles coated with silver cyanide or lead acetate have been developed for testing for the presence of hydrogen sulfide in workroom air or for other industrial hygiene purposes. These detectors or colorimetric indicators have been reviewed by Saltzman.[263] Their range of applicability is from ~1.4 to 1,100 μg/liter (1 to 800 ppm). They are suitable for roughly quantitative measurements to determine the degree of conformance with the American Conference of Governmental Industrial Hygienists (ACGIH) Threshold Limit Value of 14 μg/liter (10 ppm) for hydrogen sulfide in the air of the workplace.

Reviews of the analysis of gaseous pollutants, including hydrogen sulfide, have been published by Katz,[162, 163] Jacobs,[148] Ruch,[261] and Leithe.[187] The following description of analytical procedures contains information on standard or recommended accurate methods that are suitable for quantitative determinations of low concentrations of hydrogen sulfide in air or water.

METHYLENE BLUE METHOD (INTERSOCIETY COMMITTEE)

This method, which has been studied by Jacobs *et al.*,[150] Bamesberger and Adams,[17] and Bostrom,[41] involves absorption of the hydrogen sulfide in the measured air sample by aspiration through an alkaline suspension of cadmium hydroxide. A complete, detailed description has been published by the American Public Health Association.[305] The sulfide is precipitated as cadmium sulfide. This prevents air oxidation of the sulfide that can occur rapidly in an aqueous alkaline suspension. To minimize photodecomposition of the precipitated cadmium sulfide, STRactan 10 is added to the cadmium hydroxide slurry before sampling.[17] The collected sulfide is subsequently determined by spectrophotometric measurement of the methylene blue produced by the reaction of the sulfide with a strongly acid solution of *N*,*N*-dimethyl-*p*-phenylenediamine and ferric chloride.

Sensitivity and Range

This method is intended to provide a measurement of hydrogen sulfide in the range of 0.001 to 0.1 μg/liter. For concentrations above 0.07 μg/liter the sampling period can be reduced or the liquid volume increased either before or after aspirating. When sampling air at the maximum recommended rate of 1.5 liters/min for 2 hr, the minimum detectable sulfide concentration is 0.001 μg/liter (1.1 $\mu g/m^3$), at 760 torr and 25 C, in 10 ml of absorbing solution.

Interferences

The methylene blue reaction is highly specific for sulfide at the low concentrations usually encountered in ambient air. Strong reducing agents (e.g., sulfur dioxide) inhibit color development. Even solutions containing several micrograms of sulfide per milliliter show this effect and must be diluted to eliminate color inhibition. If sulfur dioxide is absorbed to give a sulfite concentration in excess of 10 μg/ml, color formation is retarded. Up to 40 μg/ml of this interference, however, can be overcome by adding 2 to 6 drops (0.5 ml/drop) of ferric chloride instead of a single drop for color development, and extending the reaction time to 50 min.

Nitrogen dioxide gives a pale yellow color with the sulfide reagents at 0.5 μg/ml or more. No interference is encountered when 0.4 μg/liter (0.3 ppm) of nitrogen dioxide is aspirated through a midget impinger containing a slurry of cadmium hydroxide, cadmium sulfide, and STRactan 10. If hydrogen sulfide and nitrogen dioxide are simultaneously aspirated through a cadmium hydroxide and STRactan 10 slurry, lower hydrogen sulfide results are obtained, probably because of gas phase oxidation of the hydrogen sulfide before precipitation as cadmium sulfide. Ozone at 111 $\mu g/m^3$ (57 ppb) can reduce by 15% the recovery of sulfide previously precipitated as cadmium sulfide. Sulfides in solution are oxidized by oxygen from the atmosphere unless inhibitors such as cadmium and STRactan 10 are present.

Substitution of other cation precipitants for the cadmium in the absorbent (e.g., zinc, mercury, etc.) will shift or eliminate the absorbance maximum of the solution upon addition of the acid-amine reagent.

Cadmium sulfide decomposes significantly when exposed to light unless protected by the addition of 1% STRactan to the absorbing solution before sampling.

Precision and Accuracy

A relative standard deviation of 5.3% and a recovery of 80% have been established with hydrogen sulfide permeation tubes.[17]

Apparatus

Absorber Absorbance is accomplished by a midget impinger.

Air pump The air pump is connected to a flow meter and/or gas meter having a minimum capacity of 2 liters/min through a midget impinger.

Colorimeter Colorimetry measurements are made with a red filter or spectrophotometer at 670 nm.

Air volume measurement The air meter must be capable of measuring the air flow within ±2%. Either a wet- or dry-gas meter, with contacts on the 10-liter dial or the liter and cubic-foot dial to record air volume, or a specially calibrated rotameter can be used satisfactorily. Instead of these, calibrated hypodermic needles may be used as critical orifices if the pump is capable of maintaining >0.7 atm pressure differential across the needle.[201]

Reagents

Reagents must be American Chemical Society (ACS) analytic reagent quality. They should be refrigerated when not in use. Distilled water should conform to the American Society for Testing and Materials (ASTM) Standards for Referee Reagent Water.

Amine-sulfuric acid stock solution Add 50 ml of concentrated sulfuric acid to 30 ml of water and cool. Dissolve 12 g of *N*,*N*-dimethyl-*p*-phenylenediamine, dihydrochloride, (*p*-aminodimethylaniline) (redistilled if necessary) in the acid. Do not dilute. The stock solution may be stored indefinitely under refrigeration.

Amine test solution Dilute 25 ml of the stock solution to 1 liter with 1:1 sulfuric acid.

Ferric chloride solution Dissolve 100 g of ferric chloride ($FeCl_3 \cdot 6H_2O$) in water and dilute to 100 ml.

Ammonium phosphate solution Dissolve 400 g of diammonium phosphate in water and dilute to 1 liter.

STRactan 10 (**Arabinogalactan**)

Absorbing solution Dissolve 4.3 g of cadmium sulfate ($3CdSO_4 \cdot 8H_2O$) and 0.3 g of sodium hydroxide in separate portions of water and mix. Add 10 g of STRactan 10 and dilute to 1 liter. Shake the resultant suspension vigorously before removing each aliquot. The STRactan cadmium hydroxide mixture should be freshly prepared. The solution is stable for only 3 to 5 days.

Hydrogen sulfide permeation tube Prepare or purchase a triple walled or thick-walled Teflon permeation tube[238, 239, 270] that delivers hydrogen sulfide at a maximum rate of approximately 0.1 μg/min at 25 C. This loss rate will produce a standard atmosphere containing 50 μg/m³ (36 ppb) of hydrogen sulfide when the tube is swept with a 2 liter/min airflow. Tubes having hydrogen sulfide permeation rates in the range of 0.004 to 0.33 μg/min will produce standard air concentrations in the realistic range of 1 to 90 μg/m³ of hydrogen sulfide with an airflow of 1.5 liters/min.

Concentrated, standard sulfide solution Transfer freshly boiled and cooled 0.1 M sodium hydroxide to a 1-liter volumetric flask. Flush with nitrogen to remove oxygen and adjust to volume. (Commercially available, compressed nitrogen contains trace quantities of oxygen in sufficient concentration to oxidize the small concentrations of sulfide contained in the standard and dilute standard sulfide solutions. Trace quantities of oxygen should be removed by passing the stream of tank nitrogen through a Pyrex or quartz tube containing copper turnings heated to between 400 and 450 C.) Immediately stopper the flask with a serum cap. Inject 30 ml of hydrogen sulfide gas through the septum. Shake the flask. Withdraw measured volumes of standard solution with a 10-ml hypodermic syringe and fill the resulting void with an equal volume of nitrogen. Standardize with standard iodine and thiosulfate solution in an iodine flask under a nitrogen atmosphere to minimize air oxidation. The approximate concentration of the sulfide will be 440 μg/ml of solution. The exact concentration must be determined by iodine-thiosulfate standardization immediately before dilution.

To obtain the most accurate results in the iodometric determination of sulfide in aqueous solution, the following general procedure is recommended: replace the oxygen from the flask with an inert gas such as carbon dioxide or nitrogen and add an excess of standard iodine, acidification, and back titration with standard thiosulfate and starch indicator.[172]

Diluted standard sulfide solution Dilute 10 ml of the concentrated sulfide solution to 1 liter with freshly boiled, distilled water. Protect the boiled water under a nitrogen atmosphere while cooling. Transfer the deoxygenated water to a flask previously purged with nitrogen and immediately stopper the flask. Because this sulfide solution is unstable, it should be prepared immediately prior to use. The concentration of sulfide should be approximately 4 μg/ml of solution.

Procedure

Collection of sample Aspirate the air sample through 10 ml of the absorbing solution in a midget impinger at 1.5 liters/min for a selected period up to 2 hr. The addition of 5 ml of 95% ethanol to the absorbing solution just prior to aspiration controls foaming for 2 hr (induced by the presence of STRactan 10). In addition, one or two Teflon demister discs may be slipped up over the impinger air inlet tube to a height approximately 1.25 cm from the top of the tube.

Analysis Add 1.5 ml of the amine test solution to the midget impinger through the air inlet tube and mix. Add 1 drop of ferric chloride solution and mix. (Note: See section on interferences if sulfur dioxide exceeds 10 μg/ml in the absorbing media.) Transfer the solution of a 25-ml volumetric flask. Discharge the color due to the ferric ion by adding one drop of ammonium phosphate solution. If the yellow color is not destroyed by one drop of ammonium phosphate solution, continue adding drops, one by one, until solution is decolorized. Make up to 25-ml volume with distilled water and allow to stand for 30 min. Prepare a zero reference solution in the same manner using 10 ml of unaspirated absorbing solution. Measure the absorbance of the color at 670 nm in a spectrophotometer or colorimeter set at 100% transmission against the zero reference.

Calibration

Aqueous sulfide Place 10 ml of the absorbing solution in each of a series of 25-ml volumetric flasks. Add the diluted standard sulfide solution, equivalent to 1, 2, 3, 4, and 5 μg of hydrogen sulfide, to the different flasks. Add 1.5 ml of amine-acid test solution to each flask, mix, and add 1 drop of ferric chloride solution to each flask. Mix, make up to 25 ml volume, and allow to stand for 30 min. Determine the absorbance in a spectrophotometer at 670 nm, against the sulfide-free reference solution. Prepare a standard curve of absorbance versus micrograms of hydrogen sulfide per milliliter.

Gaseous sulfide Commercially available permeation tubes containing liquefied hydrogen sulfide may be used to prepare calibration curves for use at the upper range of atmospheric concentration. Preferably the tubes should deliver hydrogen sulfide within a loss rate range of 0.003 to 0.28 μg/min. This will provide realistic concentrations of hydrogen sulfide [0.0015 to 0.139 μg/liter (1.1 to 100 ppb)] without resorting to a dilution system for the concentrations needed to determine the collection efficiency of midget impingers. Analyses of these known concentrations give calibration curves that simulate all of the operational conditions performed during the sampling and chemical procedure. This calibration curve includes the important correction for collection efficiency at various concentrations of hydrogen sulfide.

Prepare or obtain a Teflon permeation tube that emits hydrogen sulfide at a rate of 0.1 to 0.2 μg/min (0.07 to 0.14 μl/min at standard conditions of 25 C and 1 atm). A permeation tube with an effective length of 2 to 3 cm and a wall thickness of 0.318 cm will yield the desired permeation rate if held at a constant temperature of 25 $\pm$ 0.1 C. Permeation tubes containing hydrogen sulfide are calibrated under a stream of dry nitrogen to prevent the precipitation of sulfur in the walls of the tube.

To prepare standard concentrations of hydrogen sulfide, assemble the apparatus consisting of a water-cooled condenser, constant temperature bath maintained at 25 $\pm$ 0.1 C, cylinders containing pure dry nitrogen and pure dry air with appropriate pressure regulators, and needle valves and flow meters for the nitrogen and dry air diluent streams. The diluent gases are brought to temperature by passage through a 2-m long copper coil immersed in the water bath. Insert a calibrated permeation tube into the central tube of the condenser, which is maintained at the selected constant temperature by circulating water from the constant-temperature bath. Pass a stream of nitrogen over the tube at a fixed rate of approximately 50 ml/min. Dilute this gas stream to obtain the desired concentration by varying the flow rate of the clean, dry air. This flow rate can normally be varied from 0.2 to 15 liters/min. The flow rate of the sampling system determines the lower limit for the flow rate of the diluent gases. The flow rates of the nitrogen and the diluent air must be measured to an accuracy of 1% to 2%. With a tube permeating hydrogen sulfide at a rate of 0.1 μg/min, the range of concentration of hydrogen sulfide will be between 0.006 and 0.40 $\mu g/m^3$ (4 to 290 ppb), a generally satisfactory range for ambient air conditions. When higher concentrations are desired, calibrate and use longer permeation tubes.

Obviously one can prepare a multitude of simulated calibration curves by selecting different combinations of sampling rate and sampling time. Following is a description of a typical procedure for ambient air sampling of short duration, with a brief mention of a modification for 24-hr sampling. The system is designed to provide an accurate measure of hydrogen sulfide in the 0.0014 to 0.084 μg/liter (1 to 60 ppb) range. It can be easily modified to meet special needs.

The dynamic range of the colorimetric procedure fixes the total volume of the sample at 186 liters; then, to obtain linearity between the absorbance of the solution and the concentration of hydrogen sulfide in parts per million, select a constant sampling time. This fixing of the sampling time is desirable also from a practical standpoint. In this case, the sampling time is desirable also from a practical standpoint. In this case, the sampling time is 120 min. To obtain a 186-liter sample of air, a flow rate of 1.55 liter/min is required. The concentration of standard hydrogen sulfide in air is computed as follows:

$$C = \frac{Pr \times M}{R + r}$$

where C = concentration of hydrogen sulfide in parts per million, or

$$C = \frac{Pr}{R + r}$$

where C = concentration of hydrogen sulfide in micrograms per liter, Pr = permeation rate, micrograms per min, M = reciprocal of vapor density, 0.719 microliters per microgram of hydrogen sulfide, R = flow rate of diluent air, liters per minute, and r = flow rate of diluent nitrogen, liters per min. Data for a typical calibration curve are listed in Table I-1.

Table I-1. Typical calibration data

Concentrations of hydrogen sulfide (ppb)	Amount of hydrogen sulfide (μl/186 liters)	Absorbance of sample
1	0.144	0.010
5	0.795	0.056
10	1.44	0.102
20	2.88	0.205
30	4.32	0.307
40	5.76	0.410
50	7.95	0.512
60	8.64	0.615

A plot of the concentration of hydrogen sulfide in parts per million (x-axis) against absorbance of the final solution (y-axis) will yield a straight line. The reciprocal of the slope is the factor for converting absorbance to parts per million. This factor includes the correction for collection efficiency. Any deviation from the linearity at the lower concentration range indicates a change in collection efficiency of the sampling system. If the range of interest is below the dynamic range of the method, the total volume of air collected should be increased to obtain sufficient color within the dynamic range of the colorimetric procedure. Also, once the calibration factor has been established under simulated conditions, the conditions can be modified so that the concentration of hydrogen sulfide is a simple multiple of the absorbance of the colored solution.

For 24-hr sampling, the conditions can be fixed to collect 1,200 liters of sample in a larger volume of STRactan 10-cadmium hydroxide. For example, for 24 hr at 0.83 liter/min, approximately 1,200 liters of air are scrubbed. An aliquot representing 0.1 of the entire amount of sample is taken for the analysis.

The remainder of the analytic procedure is the same as described above.

The permeation tubes must be stored in a wide-mouth glass bottle containing silica gel and solid sodium hydroxide to remove moisture and hydrogen sulfide. The storage bottle is immersed to two-thirds its depth in a water bath whose temperature is kept constant at 25 $\pm$ 0.1 C.

Periodically (every 2 weeks or oftener), the permeation tubes are removed and rapidly weighed on a semimicro balance (sensitivity $\pm$ 0.01 mg) and then returned to the storage bottle. The weight loss is recorded. The tubes are ready for use when the rate of weight loss becomes constant (within $\pm$2%).

Calculation

Determine the sample volume in liters from a gas meter or from flow meter readings and sampling time. Adjust volume to 760 torr and 25 C (V_S).

$$H_2S = \frac{\mu g \times 10^3}{V_S,\ \text{liter}} = \mu g/m^3$$

Effect of Light and Storage

Hydrogen sulfide is readily volatilized from aqueous solution when pH is below 7.0. Alkaline, aqueous sulfide solutions are very unstable because the sulfide is rapidly oxidized by exposure to the air. Therefore, the dilute, alkaline sulfide standard solution must be carefully prepared under a nitrogen atmosphere. The preparation of a standard curve should be completed immediately upon dilution of the concentrated standard sulfide solution. Aqueous sulfide standard solutions may be protected from air oxidation by the addition of 0.1 M ascorbic acid.[39] Ascorbic acid should be used only in solutions that are to be analyzed by titration. Ascorbic acid interferes with the development of the methylene blue color.

Cadmium sulfide is not appreciably oxidized even when aspirated with pure oxygen in the dark. However, exposure of an impinger containing cadmium sulfide to laboratory light or to more intense light sources produces an immediate and variable photodecomposition. Losses of 50 to 90% of added sulfide have been reported routinely by a number of laboratories. Even though the addition of STRactan 10 to the absorbing solution controls the photodecomposition, it is necessary to protect the impinger from light at all times by using low actinic glass impingers, paint on the exterior impingers, or an aluminum foil wrapping.

OTHER METHYLENE BLUE METHODS

Workplace Air

A procedure that is essentially similar to the Intersociety Committee Method (above) has been adopted for hydrogen sulfide in air of the workplace by the Physical and Chemical Analysis Branch of the U.S. National Institute for Occupational Safety and Health (NIOSH). It is intended to cover the range of 0.01 to 70 μg/liter (0.008 to 50 ppm).[321]

Water Analysis

The methylene blue method is a standard technique used to determine sulfide in water and wastewater,[304] which results from the microbial decomposition of organic matter under anaerobic conditions and from certain industrial operations. Three forms of sulfide—total sulfide, dissolved sulfide, and un-ionized hydrogen sulfide—may be detected by analysis of sewage and wastewaters. Total sulfide includes the dissolved hydrogen sulfide and un-ionized hydrogen sulfide, as well as acid-soluble metallic sulfides that are present in the suspended matter. All three forms of sulfide may be determined by the methylene blue method.

Two methylene blue methods can be used to determine sulfides in water samples.[304] A drop-counting colorimetric matching method is selected when convenience rather than maximum accuracy is desired. It is effective for sulfide concentrations in the range of 0.05 to 20 mg/liter. In the more accurate method, the methylene blue color is measured with a spectrophotometer or filter photometer at 600 nm, with provision for a light path of 1 cm or longer. The range of this method is from 0.02 to 20 mg/liter of sulfide.

Samples must be collected with a minimum of aeration to prevent volatilization or oxidation of sulfide. If total sulfide only is to be determined, the samples may be preserved by adding zinc acetate solution to precipitate the sulfide as zinc sulfide. Determination of dissolved sulfide and analysis of samples that are not preserved with zinc acetate must begin within 3 min of the time of sampling. Samples to be used for determination of total sulfide must contain a representative proportion of suspended solids.

Visual Color-Matching Method

The colorimetric method is based on the reaction in which *p*-aminodimethylaniline, ferric chloride, and sulfide ion react under suitable conditions to form methylene blue. Before the color comparison is made, ammonium phosphate should be added to remove any color due to the presence of the ferric ion.

Interferences Some strong reducing agents prevent the formation of the color or diminish its intensity. High sulfide concentrations—several hundred milligrams per liter—may completely inhibit the reaction, but dilution of the sample prior to analysis eliminates this problem. Sulfite up to 10 mg/liter of sulfur dioxide has no effect, although higher concentrations retard the reaction. Thiosulfate concentrations below 10 mg/liter do not interfere seriously, but higher concentrations prevent color formation unless the thiosulfate is oxidized. The interference of sulfite and thiosulfate up to 40 mg/liter of sulfur dioxide or thiosulfate can be eliminated by increasing the amount of ferric chloride solution that is added from 2 to 6 drops and extending the reaction time to 5 min. If present, sodium hydrosulfite will interfere by releasing some sulfide when the sample is acidified. Nitrite gives a pale yellow color at concentrations as low as 0.5 mg/liter of nitrogen dioxide.

But, since nitrite and sulfide are not likely to be found together, this possible interference is of little practical importance. To eliminate a slight interfacing color due to the reagent, which may be noticeable at sulfide concentrations below 0.1 mg/liter, a dilute amine-sulfuric acid test solution is specified for concentrations of that order.

Apparatus

Matched Test Tubes Tubes approximately 125 mm long and 15 mm O.D. are the most convenient for field use. Fifty-milliliter Nessler tubes, with a corresponding increase in the amounts of sample and reagents, may be used to give a intense color to the colored solutions and, therefore, an increased sensitivity.

Droppers Droppers should deliver 20 drops per milliliter of the methylene blue solution. To secure accurate results when measuring by drops, it is essential to hold the dropper in a vertical position and to allow the drops to form slowly, so that the outside of the dropper is thoroughly drained before the drop falls.

Glass-Stoppered Bottles (Capacity: 100 to 300 ml) A biochemical oxygen demand (BOD) incubation bottle is recommended because its stopper is ground in such a way that it minimizes the possibility of entrapping air, and its especially designed lip provides a water seal.

Reagents

Zinc Acetate Solution (2 N) Dissolve 220 g of zinc acetate ($Zn[C_2H_3O_2]_2 \cdot 2H_2O$) in 870 ml of water to make 1 liter of solution.

Sodium Carbonate Solution Dissolve 5.0 g of sodium carbonate in distilled water and dilute to 100 ml.

Amine-Sulfuric Acid Stock Reagent Dissolve 26.6 g of *N,N*-dimethyl-*p*-phenylenediamine oxalate (also called *p*-aminodimethylaniline oxalate) in a cold mixture of 50 ml of concentrated sulfuric acid and 20 ml of distilled water. Cool, then dilute to 100 ml with distilled water. Store in a dark glass bottle. This stock solution may discolor on aging, but its usefulness is unimpaired.

Amine-Sulfuric Acid Reagent Dilute 25 ml of amine-sulfuric acid stock solution with 975 ml of 1 + 1 sulfuric acid. Store in a dark glass bottle.

Ferric Chloride Solution Dissolve 100 g of ferric chloride ($FeCl_3 \cdot 6H_2O$) in 39 ml of water. This makes 100 ml of solution.

Sulfuric Acid Solution, 1 + 1 Add, cautiously, 500 ml of concentrated sulfuric acid to 500 ml of distilled water, continuously mixing. Cool the solution before using.

Diammonium Hydrogen Phosphate Solution Dissolve 40 g of dibasic ammonium phosphate in distilled water and dilute to 100 ml.

Stock Sulfide Solution Dissolve 4.10 g of sodium sulfide trihydrate ($Na_2S \cdot 3H_2O$) in boiled, cooled distilled water. Weight the sodium sulfide from a well-stoppered weighing bottle. Dilute to 1 liter in a volumetric flask to form a solution containing 1.0 mg of sulfur/1.0 ml of solution. If the weight of sodium sulfide trihydrate used is other than that recommended, calculate the sulfide concentration as follows:

$$\text{mg/liter of sulfur} = 242.8 \times B$$

where B = grams of sodium sulfide trihydrate per liter. Prepare the stock solution daily.

Standard Sulfide Solution Take 20.0 ml of stock solution or an appropriate aliquot which contains 20.0 mg of sulfide. Dilute to 1 liter with boiled, cooled, distilled water. Because of its instability, prepare this solution as needed. Standardize by pipetting 100 ml of solution into an Erlenmeyer flask and immediately add 10.00 ml of standard 0.0250 *N* iodine solution and 2 drops of concentrated hydrochloric acid. Titrate the residual iodine with standard 0.0250 *N* sodium thiosulfate titrant, using a starch indicator at the end point. Run a blank on the reagents. Calculate the sulfide concentration, which should be approximately 20 mg/liter of sulfur or 1 ml = 20 μg, as follows:

$$\text{mg/liter of sulfur} = (10.00 - C - D) \times 4$$

where C = milliliters of 0.0250 *N* sodium thiosulfate titrant required for titration, and D = milliliters of 0.0250 *N* iodine solution used for reagent blank.

Methylene Blue Solution I Use the U.S. Pharmacopeia (USP) grade of the dye, or one that has been certified by the Biological Stain Commission. The percentage of actual dye content, which should be reported on the label, should be 84% or more. Dissolve 1.0 g of methylene blue powder in enough water to make 1 liter. This solution will be approximately the correct strength, but because of variation between different lots of dye, it must be standardized against sulfide solutions of known strength and its concentration adjusted so that 1 drop (0.05 ml) of solution will be equivalent to 1.0 mg/liter of sulfide.

Standardization Determine the number of drops of methylene blue solution that will produce a color equivalent to that obtained with a measured aliquot of the standard sulfide solution in accordance with the procedure described on p. 130 under "Color Development." After making this analysis, adjust the methylene blue solution either by diluting with water or by adding more dye so that 1 drop is equivalent to 1.0 mg/liter of sulfide. After making an adjustment, repeat the colorimetric determination to check the adjusted solution. The methylene blue solution is stable for a year if kept in the dark and tightly stoppered.

Methylene Blue Solution II Dilute 10.00 ml of the adjusted methylene blue solution I to 100 ml, making one drop (0.05 ml) equivalent to 0.1 mg/liter of sulfide.

Sodium Hydroxide 6 N

Procedure for total sulfide

Sample Pretreatment Add 3 or 4 drops of zinc acetate solution to a 100-ml sample, then add a few drops of sodium carbonate solution. After allowing the precipitated zinc sulfide to settle, decant the clear liquid. Add sufficient water to the precipitated slurry to restore the volume to 100 ml. When interferences are absent, the pretreatment may be omitted.

Color Development Fill two color comparison tubes to the 7.5 ml mark with sample. Add to one tube 0.5 ml of amine-sulfuric acid reagent and 3 drops (0.15 ml) of ferric chloride solution; stopper and mix the contents immediately by inverting the tube slowly, only once. Add to the other tube 0.5 ml of 1 + 1 sulfuric acid and 3 drops (0.15 ml) of ferric chloride solution; stopper and mix the contents immediately by inverting the tube slowly, only once.

The presence of sulfide ions will be indicated by the immediate appearance of blue color in the first tube. Complete color development requires about 1 min. One to 5 min after the color first appears, add 1.6 ml of dibasic ammonium phosphate solution to each tube.

Visual Color Estimation Add methylene solution I or II, depending on the sulfide concentration and the desired accuracy of the test. Drop by drop, add the solution to the contents of the second tube until the color imparted by the methylene blue matches that developed in the first tube. Record the total number of drops of methylene blue solution added to the contents of the second tube.

Procedure for dissolved sulfides Fill a glass-stoppered BOD bottle with the sample and eliminate air bubbles. Add 0.5 ml of 6 *N* aluminum chloride solution and 0.5 ml of 6 *N* sodium hydroxide. Stopper the bottle and flocculate the precipitate by rotating the bottle back and forth about a transverse axis. Allow the floc to settle. Proceed with the clear supernatant liquid as directed under "Color Development," above.

Procedure for un-ionized hydrogen sulfide Determine the pH of the original sample. Calculate the concentration of un-ionized hydrogen sulfide by multiplying the concentration of dissolved sulfide by a suitable factor as given in Table I-2. These factors are applicable at a temperature of 25 C.

Table I-2. Hydrogen sulfide factors[a]

pH	Factor	pH	Factor	pH	Factor
5.0	0.99	6.7	0.61	7.4	0.24
5.4	0.97	6.8	0.55	7.5	0.20
5.8	0.92	6.9	0.49	7.6	0.16
6.0	0.89	7.0	0.44	7.7	0.13
6.2	0.83	7.1	0.38	7.8	0.11
6.4	0.76	7.2	0.33	7.9	0.089
6.5	0.71	7.3	0.28	8.0	0.072
6.6	0.66				

[a] Based on: $K_1 = 1.1 \times 10^{-7}$ (25 C); ionic strength, μ = 0.02.

Calculation

With methylene blue solution I, adjusted so that 1 drop (0.05 ml) corresponds to 1.0 mg/liter sulfide when 7.5 ml of sample are used:

$$\text{mg/liter sulfide} = \text{no. drops} = \text{ml} \times 20$$

With methylene blue solution II, adjusted so that 1 drop (0.05 ml) corresponds to 0.1 mg/liter sulfide when 7.5 ml of sample are used:

$$\text{mg/liter sulfide} = \text{no. drops} \times 0.1 = \text{ml} \times 2$$

If dilution is necessary, mutliply the result by the appropriate factor. With care, the accuracy is about $\pm 10\%$.

Photometric Method

Apparatus Colorimetric equipment—One of the following is required:

Spectrophotometer, for use at 600 nm, providing a light path of 1 cm or longer

Filter photometer, providing a light path of 1 cm or longer and equipped with a red filter exhibiting maximum transmittance near 600 nm.

Graduated cylinders or flasks, 50-ml capacity.

Reagents All reagents listed for the Visual Color-Matching Method are required except the standard methylene blue solutions.

Procedure

Preparation of Standard Curve Add to separate 50-ml graduated cylinders or flasks the following volumes of standard sulfide solution (1.0 ml = 20 μg): 0 (reagent blank), 0.5, 1.0, 2.0, 3.0, 4.0, and 5.0 ml in order to prepare a sulfide series containing 0, 10, 20, 40, 60, 80, and 100 μg, respectively. Dilute to 50 ml with boiled and cooled distilled water.

Add 0.5 ml amine-sulfuric acid reagent and mix. Then add 2 drops (0.10 ml) of ferric chloride solution and mix again. After 1 min add 1.5 ml of diammonium hydrogen phosphate solution and mix. Measure the absorbance against the reagent blank (usually colorless) at a wavelength of 600 nm. Plot absorbance against micrograms of sulfur.

Total Sulfide See "Sample Pretreatment" on p. 130. When interference is absent, omit this step. Measure 50 ml of distilled water and 50 ml of sample (or a suitable aliquot diluted to 50 ml) into separate graduated cylinders or flasks. Complete the determination as directed under "Color Development" on p. 130 and refer to the standard curve for the sulfide concentration.

Dissolved Sulfide Remove the suspended matter in the sample and complete the determination as directed under "Procedure for Dissolved Sulfides" on p. 130.

Un-ionized Hydrogen Sulfide Determine the pH of the original sample and calculate the concentration of the un-ionized hydrogen sulfide as directed under "Procedure for Un-ionized Hydrogen Sulfide" in this chapter.

Calculation

$$\text{mg/liter sulfur} = \frac{\mu\text{g sulfur}}{\text{ml sample}}$$

Results by the photometric method are estimated to be equal to, or perhaps more reliable than, those obtained by the visual comparison method.

MOLYBDENUM BLUE METHOD

Buck and Stratmann[52] have developed an analytic method for hydrogen sulfide in air in which the gas is absorbed in an impinger containing an alkaline suspension of cadmium hydroxide. The gas is released subsequently by the addition of a solution of stannous chloride in hydrochloric acid. The liberated hydrogen sulfide is then introduced into a solution of ammonium molybdate to form molybdenum blue which is measured photometrically at 570 nm.

Apparatus

The apparatus consists of an impinger with a bottom part that can fit into a centrifuge. The nozzle opening of the impinger is 2.50 mm and when connected to a pump with a capacity of 2 to 3 m^3/hr, a constant underpressure of 140 ± 10 mm of mercury is maintained. The impinger is filled with 25 ml of a solution of 2.15 g cadmium sulfate

($3CdSO_4 \cdot 8H_2O$) in 250 ml of distilled water and 25 ml of a solution containing 1 g sodium hydroxide in 250 ml of water.

Procedure

After passing about 1 m of measured air sample into the impinger solution over a 30-min period, the suspension in the bottom part of the impinger is centrifuged. The time interval between sampling and centrifuging should not exceed 4 hr. The impinger part containing the residue is connected to an apparatus for the separation of hydrogen sulfide in a stream of nitrogen at a flow rate of 7 to 10 liter/hr. The nitrogen is purified by initially passing it through an activated carbon filter and a glass wool filter impregnated with cadmium sulfate solution. A 30-ml solution of 100 g stannous chloride ($SnCl_2 \cdot 2H_2O$) in 1 liter of 12 *N* hydrochloric acid is introduced into a reversely connected scrubber and lifted by the nitrogen stream into the vessel containing the cadmium hydroxide residue. Excess hydrochloric acid vapors are removed from the decomposition products in two scrubbers, in series, following the impinger part. Each of these two scrubbers contains 3 ml of a dilute solution of 100 g of stannous chloride in 600 ml of concentrated hydrochloric acid. The volume is brought up to 1 liter by adding water. The gas stream is then passed into a third scrubber containing 50 ml of ammonium molybdate solution made up of three parts by volume of a 3.33% aqueous ammonium molybdate solution plus two parts by volume of a solution of 0.5 g of urea in 1 liter of *N* sulfuric acid. (The ammonium molybdate solution must be renewed frequently.) Decomposition ceases after a 20-min continuous passage of nitrogen. The molybdenum solution is rinsed into a volumetric flask and allowed to stand for 20 min to complete the color reaction. The absorbance of the colored solution is measured then with a spectrophotometer at 570 nm, using a 1-cm path cell, against water. At absorbance above 1.0, an aliquot of the blue solution must be diluted.

Calibration

A calibration curve is prepared by using portions of a standard solution of 0.6 g of sodium pyrosulfite/liter. Between 0.1 and 0.8 ml of the standard solution are used to convert the hydrogen sulfide, which is liberated as described above, and to measure as molybdenum blue. The extinction obtained from the calibration curve must be multiplied by 1.07 to correct for incomplete absorption and by 1.32 to correct for the loss associated with oxidation by air.

The molar extinction coefficient for hydrogen sulfide is 10,000, the same as for molybdenum blue. According to Buck and Stratmann,[52] the detection limit of the method is 20 μg of hydrogen sulfide/m^3 of air and the relative standard deviation is $\pm 5\%$. Mercaptans, carbon disulfide, sulfur dioxide, and nitrogen dioxide in the concentrations expected in the air do not interfere in this method.

GAS CHROMATOGRAPHIC ANALYSIS—FLAME PHOTOMETER DETECTOR

This procedure for the detection and determination of low molecular weight sulfur-containing gases in the atmosphere, including hydrogen sulfide, sulfur dioxide, methyl mercaptan, and dimethyl sulfide, uses the specificity of the flame photometric detector (FPD).[146, 147]

The atmospheric Teflon sampling-line is attached to a multiport switching valve and air pump. The sampled air and calibration gas are alternately drawn through the valve sample loop. On a preset cycle the valve sample loop is switched to the carrier gas. The sample in the loop is purged into the gas chromatographic (GC) column for sample resolution. On alternate cycles the calibration gas is purged into the GC column to provide standard reference peak heights and to maintain column conditioning (priming).

The eluted sulfur compounds pass through a hydrogen flame and the sulfur is converted to the "excited" sulfur molecule which is at a higher energy state. Upon returning to the ground state, these molecules emit light of characteristic wavelengths between 300 and 425 nm.[47, 79] This light passes through a narrow-band optical filter and is detected by a photomultiplier (PM) tube. The current produced in the PM tube is amplified by an electrometer. The magnitude of the response is displayed on a potentiometric recorder or other suitable device. The analyzer is calibrated with hydrogen sulfide, sulfur dioxide, methanethiol, and dimethyl sulfide permeation tubes[238, 271] and a dual-flow gas dilution device[90] that can produce reference standard atmospheres down to the detection limits of the method.

Sensitivity and Range

The sensitivity, repeatability, and accuracy of the method are dependent upon many variables including the materials used to construct the multiport valve and the total system for handling the gas chromatograph sample; carrier and make-up gas flow rates; bias voltage; temperature and type of the PM tube; and the column preparation technique. The limits of detection at twice the noise level for hydrogen sulfide, sulfur dioxide, methyl mercaptan, and dimethyl sulfide range from 5 to 13 $\mu g/m^3$ [116, 248] without the use of sample concentration techniques such as freeze-out loops. The sensitivity of the detector, reported as mass per unit time, is 1.6×10^{-4} $\mu g/sec$ or 6.0×10^{-5} $\mu l/sec$, since this is a dynamic system.

The response of the system is nonlinear, but a linear relationship is obtained by plotting the response against concentration on a log scale or by using a log-linear amplifier. By using either of these techniques the linear dynamic range is approximately 130 to 500 $\mu g/m^3$ with a 1% noise level.

Interferences

The FPD is based on a spectroscopic principle. An optimal filter isolates the emission wavelength for the sulfur species at 394 ± 5 nm from other extraneous light sources. Other emission bands of sulfur of almost equal intensity at 374 and 383.8 nm are also suitable for the quantification of sulfur.

Phosphorus presents a potential interference at 394 ± 5 nm. Phosphorus-containing gases are not generally found in ambient air unless the samples are obtained near known sources of phosphorus-containing insecticides. Under the conditions of the method (gas chromatograhic separation) it is unlikely, however, that phosphorus-containing gases would have the same retention times as the sulfur gases. Therefore, they would not present a real source of interference. The FPD discrimination ratio is approximately 10,000:1 for hydrocarbons and at least 1,000:1 for other gases.[248]

Air produces a spectral continuum which yields a measurable signal at 394 nm and contributes to the background noise of the method. However, it is not considered to be a significant interference under the conditions of the method.

For a substance to interfere it must meet three conditions: it must emit light within the band pass of the filter; it must have an elution time very close to those of sulfur dioxide, hydrogen sulfide, methanethiol, or dimethyl sulfide; and it must be present in the sample at a concentration that is detectable by this procedure.

Precision and Accuracy

Precision or repeatability of the flame photometric detector depends on close control of the sample flow and hydrogen flow to the detector. An airflow change of 1% alters the response approximately 2%.

Reproducible peak heights are primarily dependent upon the materials of construction used in the GC, the close control of all GC operating variables, and the technique of column preparation. It is necessary to precondition (prime) the total analysis system with a series of standard injections to achieve elution equilibrium. After periods of disuse, equilibrium must again be attained by the serial injection of standard sulfur gas mixtures. In the automated, semicontinuous mode, sample air and a standard reference gas mixture should be alternately injected into the instrument on a timed cycle (typically every 5 to 10 min) to maintain column conditioning. The instrument should have the capability of receiving repetitive standard gas injections through a manual override control until equilibrium is achieved. This would be indicated by the uniformity of the resultant peak heights as recorded on the strip chart.

Repetitive sampling of standard reference gases containing 0.08 μg/liter (0.06 ppm) of hydrogen sulfide and 0.1 μg/liter (0.04 ppm) of sulfur dioxide over several 24-hr periods gave a relative standard deviation of $<3\%$ of the amount present.[299]

The accuracy of the method depends on the ability to control the flow and temperature of dilution gas over certified or calibrated permeation tubes maintained in a gas dilution device.[90]

This procedure was described by the Intersociety Committee.[146, 147] Additional information on chromatographic techniques for the analysis of trace concentrations of gaseous sulfur compounds has been presented by Koppe and Adams,[174] Hartmann,[131, 310] and Stevens *et al.*[298]

MISCELLANEOUS METHODS

A method for the electrochemical or potentiometric determination of hydrogen sulfide in air in the range of 0.7 to 70 μg/liter (0.5 to 5 ppm) was developed by Oehme and Wyden.[237] They used an indicating, selective ion electrode consisting of a silver rod coated with a layer of silver sulfide. In a modification of this technique, the Delwiche method[84] is used to determine sulfide in the range of 10 to 100 $\mu g/m^3$. This method is based on the precipitation of colloidal lead sulfide in the presence of excess lead ions by lead acetate and turbidimetric measurement of the suspension by the absorbance at 500 nm. It has been improved by the addition of a silver-sulfide selective ion electrode[178] (Orion model 94-16).

Hydrogen sulfide can be trapped in a sorption tube backed with glass beads coated with a thin layer of silver sulfate.[51] The beads (2 to 3 mm diameter) are placed on a filter. They are coated by pouring over them a mixture of equal volumes of a saturated silver sulfate solution and a 5% solution of potassium bisulfate in a clean atmosphere. After draining off the excess liquid, the moist beads are dried in a drying oven and then packed to a height of 10 cm in an absorption tube fitted with a ground-glass joint. The air sample may be passed through the tube at a rate of 3 to 4 liters/min. Trapped hydrogen sulfide may be desorbed by 25 ml of stannous chloride solution (100 g $SnCl_2{\cdot}2H_2O$ in 1 liter of concentrated hydrochloric acid). The liberated hydrogen sulfide may be analyzed by the methylene blue or molybdenum blue method.

Adams and Koppe[3] used a gas chromatograph coupled with a microcoulometric, bromine titration cell to determine hydrogen sulfide and other sulfur-containing gases emitted from kraft paper mills. Thoen *et al.*[308] have developed instrumentation involving electrometric titration for the quantitative measurement of sulfur compounds, including hydrogen sulfide in concentrations down to a lower limit of 15 $\mu g/m.^3$ Another instrument developed in Germany for measuring hydrogen sulfide employs sensitive galvanic measuring cells.[181]

The use of an ion exchange resin for field sampling of air containing hydrogen sulfide or for contaminated water samples has been described by Paez and Guagnini.[243] Desorption of the hydrogen sulfide from the resin is accomplished with sodium hydroxide solution. The resultant sulfide can be analyzed by the methylene blue method.

E. Kothny of the Air and Industrial Hygiene Laboratory at the California State Department of Public Health, Berkeley (personal communication, May 1976), has proposed a modification of the methylene blue method wherein the hydrogen sulfide or sulfide solution is treated with 1-(2-benzothiazolyl azo)-2-methyl-4-naphthol (BTN) reagent. BTN may be synthesized as follows: Dissolve 5 g of 2-hydrazinebenzothiazol (Eastman 3967) in 20 ml hot acetic acid. Dissolve 5.3 g of 2-methyl-1,4-naphthoquinone (Menadione, vitamin K_3) in 30 ml of hot acetic acid and mix with the thiazol. Heat to boiling for 5 min. Let stand until cold. Then, filter, wash with 10 ml cold acetic acid, and dry at 60 C. The yield is 55% of BTN.

BTN Stock Solution

Dissolve 100 mg (dye content$>$90%) BTN in a mixture of 50 ml acetone and 50 ml of 0.2 *N* sodium hydroxide. This stock solution is stable for 1 year in the dark.

Silver Nitrate Solution (0.01 N)

Dilute 0.1 *N* silver nitrate solution to one tenth with distilled water.

2% Protective Colloid

Either hydroxypropyl methylcellulose (Methocel HG) or gum acacia can be used. Mix 20 g with 100 ml of methanol and pour this mixture into 900 ml of distilled water. Let age overnight, stirring occasionally.

Reagent Solution

This is prepared as follows: start with 400 ml of distilled water. Add in this order: 8 ml 50% sodium hydroxide; 3 ml of 0.01 *N* silver nitrate; 10 ml of 2% protective colloid; 20 ml of BTN stock solution, 0.1%; and 100 ml of isopropanol (not necessary with gum acacia). Make up to 1,000 ml with distilled water. The color of the solution should turn to a slightly purple pink. If it turns totally pink, there is excess of silver. Add the BTN stock solution, 1 ml at a time, until a slight purple color persists. If the reagent is too purple after aging for 15 min, add 0.01 *N* silver nitrate, drop by drop, shaking after each addition. The procedure is generally not necessary. Age for 24 hr before use. Place 10 ml into an impinger. Sample the air at the rate of 0.5 to 1 liter/min. Compare in a spectrophotometer against reagent at 575 nm. The reagent is light sensitive. In the dark it keeps well for several months.

SUMMARY

This review of sampling and analytic techniques for hydrogen sulfide indicates that although a considerable variety of methods may be used for qualitative estimations, only a relatively few techniques have the sensitivity and precision to measure concentrations of interest in studies of the ambient air or odor threshold.

The most accurate laboratory method for both air and water samples appears to be that based on treatment of the sample with an alkaline suspension of cadmium hydroxide, precipitation of the hydrogen sulfide as cadmium sulfide, and subsequent reaction of the sulfide with a strongly acid solution of *N*,*N*-dimethyl-*p*-phenylenediamine and ferric chloride to form methylene blue. The resultant intensity of the color is measured by absorbance at 670 nm in a spectrophotometer. The methylene blue reaction is highly specific for sulfide at the low concentrations usually present in ambient air.

Low molecular weight, sulfur-containing gases in the air may be determined by gas chromatographic analysis using a hydrogen flame photometric detector. This method can be implemented with automatic instrumentation containing a multiport switching valve operated on a preset cycle, a calibrated air pump, a gas chromatograph equipped with a flame photometric detector, narrow band optical filter, PM tube, electrometer, and potentiometric recorder. The analyzer can be calibrated to detect and measure hydrogen sulfide, sulfur dioxide, methyl mercaptan, and dimethyl sulfide down to a lower limit of 5 to 13 $\mu g/m^3$.

Roughly quantitative field air sampling and measuring techniques include paper tape or tile that is impregnated with lead acetate and exposed for various periods at selected locations. The extent of staining by hydrogen sulfide is measured by light transmission or reflectance. An automatic paper tape sampler based on this principle has been developed for field use. The impregnated lead acetate tape is light-sensitive, however. Mercuric chloride, which has been proposed as a substitute for lead acetate in impregnated paper tape, is more resistant to fading by light. However, the presence of sulfur dioxide in the air causes interference in the hydrogen sulfide measurement.

The silver-sulfide selective ion electrode has been employed in the electrochemical or potentiometric determination of hydrogen sulfide.

Another modification makes use of a gas chromatograph coupled with a microcoulometric bromine titration cell to determine hydrogen sulfide and other sulfur-containing gases. Hydrogen sulfide from the air sample can be trapped by a sorption tube containing glass beads coated with a thin layer of silver sulfate. Alternatively, the use of an ion exchange resin has been proposed to remove hydrogen sulfide from contaminated air or water samples. The trapped hydrogen sulfide may be desorbed from the coated glass beads by a solution of stannous chloride in concentrated hydrochloric acid, from the ion exchange resin, or by a sodium hydroxide solution. In either case, the liberated sulfide may be analyzed by the methylene blue method.

Appendix II

Hydrogen Sulphide Literature*

By **C. W. Mitchell,** Passed Assistant Surgeon, United States Public Health Service, and **S. J. Davenport,** Translator, Bureau of Mines, Department of the Interior

In connection with a recent investigation of hydrogen sulphide poisoning, a study was made of the literature dealing with this problem. A paucity of articles on this subject seemed to be indicated by a superficial examination, but upon a more careful and detailed study a number of excellent reports were found. It was difficult to locate a few of the more important ones. Some, referred to in early literature on this subject, could not be found, although allusion to them appeared in other articles. For the convenience of those engaged in the future study of hydrogen sulphide poisoning, a list has been made of the most important articles on this subject. In addition, a resume is given of the contributions of the principal workers.

The observations relative to hyrogen sulphide poisoning may be divided into three groups: Those which deal with the actual cases of poisoning; those which concern experimental pharmacological study of animals subjected to the poison; and those relating to chemical effects of the poison on the blood.

Hydrogen sulphide gas was known to the ancients and has been described as "sulphurous vapor" and as "divine water," its name being taken from the Greek word theion, meaning divine or sulphurous. The gas was first examined by Rouelle[1] in 1773, but Scheele[2] in 1777 was the first to make a systematic study of the compound, and we owe much of our knowledge to his work.

About this time there occurred in Paris numerous accidental deaths due to the gases from the sewers. A commission was appointed to investigate the conditions, and in 1785 M. Hallé[3] reported the results of the study. These early workers recognized two types of poisoning which they thought were quite distinct. One they termed the "mitte," which was described as an inflammation of the eyes and mucous membranes, and the other, called the "plomb," was described as a type of asphyxia. They did not understand that hydrogen sulphide was the cause of the poisoning.

* Work done in cooperation with the Bureau of Mines, Department of the Interior.

This paper appeared in the January 4, 1924 issue of *Public Health Reports* 30(1):1–13. It is reprinted here in its entirety.

During the next few years Dupuytren, Thenard, and Barruel[4] by chemical analyses proved the presence of hydrogen sulphide in the sewers, and this gas was associated with the accidents and believed to be the cause of many of them.

In 1803 Chaussier[5] described the first animal experimental study, the records of which are available. He stated that the effects of breathing hydrogen sulphide were well known, probably referring to the work of Hallé 18 years previous. Chaussier's experiments indicated that poisoning might occur from surface absorption. He found that an animal would die in from 20 to 30 minutes if the body were exposed to the gas, although the animal was allowed to breathe fresh air. When he injected quantities of gas into the rectum of animals, and also into the stomach, symptoms of poisoning appeared and death resulted. Nine liters of gas injected into the rectum of a horse caused death within one minute.

Shortly after Chaussier's experiments, Nysten[6] injected hydrogen sulphide solution into the veins of experimental animals. Three injections of hydrogen sulphide, 10 c. c. of a saturated solution, in a dog caused the following symptoms:

1. Animal became excited and breathed deeply.
2. Made convulsive movements but later became calm.
3. Suffered from asphyxia—respirations feeble and slow, and the animal appeared as though dead. The following day, however, the animal was normal and apparently not affected by these injections.

Nysten concluded that the animal would probably not have been able to resist this quantity of hydrogen sulphide if it had been diluted in 500 to 600 volumes of air and given through the lungs (even though diluted in 500 to 600 volumes of air.) About this time Thenard and Dupuytren[7] also began to experiment with hydrogen sulphide. It is believed that the recognition of hydrogen sulphide as a cause of the accidents in the sewers was due to Dupuytren's experiments. In their experiments, Thenard and Dupuytren found that 0.066 per cent of hydrogen sulphide was fatal for a greenfinch, 0.125 fatal for dogs, and 0.4 fatal for a horse. The records of these early experiments were first available in 1812, but the experiments were probably completed a few years previous. In speaking of these experiments in 1827, Thenard gives the priority to Chaussier. Thenard also mentions Magendie[8] as having injected hydrogen sulphide into the venous system of animals. He found that some of the gas was liberated in the lungs but that a greater part was carried in solution in the arterial blood for a certain time and that it affected the red color of the blood. This in all probability is the earliest observation to associate a change in the hemoglobin with this gas.

In 1829 another commission was appointed to investigate the Paris sewers. Parent-Duchatelet[9] submitted a comprehensive report, which included a description of the means taken to prevent accidents, such as walling off the sewer, pumping in fresh air, burning the gas, etc. Analyses of the air in the sewers were made by Gaultier de Claubry. He stated that as high as 2.99 per cent of hydrogen sulphide was present and that the mean was 2.29 per cent. Parent-Duchatelet stated that a dog might live for eight days in this atmosphere. The method of analysis used is not given, but it was concluded from these experiments that the percentages as determined by Chaussier and other early workers were too low.

By 1836 the condition of the Paris sewers evidently had not been greatly improved, for D'Arcet[10] reported the death of three young men from the gas liberated from defective sewer connections.

During the building of the tunnel under the Thames by Sir M. Brunel, hydrogen sulphide seeped through the walls and poisoned the men engaged in the work. Taylor[11] stated that the symptoms of poisoning were marked and that a number of men died. The affection ceased only when the tunnel was completed and ventilation was established.

Christison[12] in his description of hydrogen sulphide poisoning recognized that the two types of poisoning, early observed in the study of the gases of the Paris sewers, were due to hydrogen sulphide and were produced by different percentages of the gas. One type was acute poisoning, due to a high percentage of hydrogen sulphide gas in the atmosphere, while the other was subacute and due to a smaller amount of gas. He quoted the percentage as given by Thenard and Chaussier.

Apparently the first case of hydrogen sulphide poisoning reported in America is that mentioned by Bell[13] and Raphael[14] in 1851, both of whom report an accident due to gases formed and liberated in an outhouse, which was therefore comparable to the accidents of the Paris sewers. Though no analysis was made of the gas, the doctors in attendance recognized that hydrogen sulphide was the cause of the accident. The symptoms noted and the method of treatment used were described in detail. This report is instructive, particularly so as it directs attention to the severe intoxication which hydrogen sulphide may produce.

In 1857 Bernard[15] injected hydrogen sulphide solution into venous blood and proved that hydrogen sulphide was eliminated through the lungs, as determined by the blackening of lead acetate when exposed to the exhaled air. He believed that the arterial blood carried the hydrogen sulphide, which was poisonous. Bernard also found that an animal could often be revived by being given artifical respiration.

Barker[16] in 1858 recognized that hydrogen sulphide, in small quantities, first accelerated the respiration; this acceleration was soon followed by a decrease in the respiratory rate and the appearance of dyspnea. He did not state the percentage of gas which produced these symptoms, but reported that 1 part of hydrogen sulphide in 18 parts of air immediately killed birds and that dogs were asphyxiated by 1 part in

210 parts of air. He also recognized that the symptoms of hydrogen sulphide poisoning were similar to those seen in poisoning by sewer gas and that air from sewers might produce morbid symptoms due to the hydrogen sulphide in the sewer gas.

Holden and Letheby,[17] in 1861, reported the medical history of cases of poisoning in London sewers and also gave the findings of a postmortem examination. They observed that the poisoning altered the blood, for it was found to be dark and liquid even after 4 days following death. The lungs were pale, crepitant, and somewhat emphysematous. A number of dead sewer rats found near the place where the men were killed presented similar pathological findings. Holden and Letheby concluded that the hydrogen sulphide was probably formed by acid acting upon the sewer mud.

The next worker, Hoppe-Seyler,[18] in 1863 was the first to study the chemical action on blood. He observed that when hydrogen sulphide was passed through blood a dark green pigment was deposited which was similar to the greenish discoloration of cadavers. This change was thought to be due to the action of hydrogen sulphide on the oxyhemoglobin of the blood with the formation of a substance termed "sulphmethemoglobin." An absorption spectrum was found with two bands in the red, one near to C and the second about midway between C and D. These findings were confirmed by Arake,[19] who concluded that sulphmethemoglobin was a compound which might be decomposed to hemochromogen through the action of caustic soda in solution. Sulphmethemoglobin was thought to be derived from the hemoglobin of the blood.

The work of Hoppe-Seyler led to an intensive chemical study of the action of hydrogen sulphide on blood, special attention being given to its action on the hemoglobin. This study led to the discovery of a disease termed "sulphemoglobinaemia."

The conclusion of Gamgee[20] that blood previously treated with carbon dioxide is not decomposed by hydrogen sulphide agrees with the findings of Hoppe-Seyler and was later confirmed by Lewisson[21] and Kuhne.[22] Gamgee did not believe, however, that there was sufficient evidence to support the theory of the existence of a special compound, sulphmethemoglobin, or that it explained the spectrum which has been described. He reasoned that there was a mixture of decomposition products of oxyhemoglobin brought about by the action of hydrogen sulphide upon blood and that it was those products which produced the absorption bands.

Laborde[23] in 1886 found by repeated spectroscopic examination of the blood that injections of hydrogen sulphide solution into the veins were followed by changes in the spectroscopic bands similar to those produced by the action of hydrogen sulphide on hemoglobin. He concluded that hydrogen sulphide was carried to the central nervous system, for in an examination of specimens of brain tissue preserved in alcohol he found a change in the vascular system. There was also changes in the organic substance of the respiratory center. He believed that hydrogen sulphide had a direct action upon the respiratory center, the vagus nerves, and the hemoglobin.

In 1898 Harnack[24] demonstrated that when hemoglobin was made oxygen-free by saturation with carbon dioxide, as described by Hoppe-Seyler and others, hydrogen sulphide had no action, but that if the blood were not so saturated with carbon dioxide, the dark red color with characteristic absorption bands was formed. This spectrum consisted of a band between C and D, extending from λ610 to λ625. Further, a decomposition of the blood-coloring matter occurred when oxygen was present. Acid hemoglobin was formed and hematin might occur in rare cases.

Clarke and Hurtley,[25] in 1907, produced a compound soluble in aqueous solution which they termed sulphemoglobin and believed to be the same as that described by Harnack. This compound was characterized by the production of a purple color and by the development of an absorption band in the red region of the spectrum from λ610 to λ625. It formed quickly and readily by adding hydrogen sulphide to blood or by adding a solution of hydrogen sulphide to defibrinated and laked blood. The solution was remarkably stable, but readily changed to acid hematin by the addition of a small quantity of acid. The band in the red was not affected by ammonia or ammonium sulphide. Clarke found that this substance was produced with minute quantities of hydrogen sulphide within a period of 25 minutes, but in the presence of phenyl hydrazine within 3 seconds. From this experiment Clarke and Hurtley suggested the theory that the presence of a powerful reducing agent in the blood would allow a mere trace of hydrogen sulphide to act on the blood, resulting in the formation of this compound, called sulphemoglobin.

Van der Beigh,[26] two years before, had demonstrated that certain organisms isolated from the stool of patients suffering from constipation formed hydrogen sulphide, and he believed that these organisms were capable of bringing about, in the human body, a transformation of the hemoglobin to sulphemoglobin. He was the first to recognize that sulphemoglobinaemia was a distinct disease.

West and Clarke,[27] while studying a case of sulphemoglobinaemia, confirmed the theory advanced by Clarke and Hurtley that in cases of this disease very small amounts of hydrogen sulphide will combine with the hemoglobin to form sulphemoglobin. They found that hydrogen sulphide in high dilution combined with blood. The dilutions were such that hydrogen sulphide could not be detected by chemical means.

It remained for Wallis[28] to find that blood from a patient suffering from sulphemoglobinaemia quickly reduced normal blood, the former containing a powerful reducing substance. This reducing agent is probably a hydroxylamine derivative thought to be produced by a nitrobacillus which inhabits the buccal cavity. Sulphemoglobin is present in these pathological cases as a constituent of the blood, and the existence of this compound depends upon two factors: i.e., the production of a powerful reducing agent and the production of hydrogen sulphide, thought to be formed in the gastrointestinal tract. In cases of hydrogen sulphide poisoning, however, sulphemoglobin may not be found. The fact that hydrogen sulphide acts upon the hemoglobin with the formation of sulphemoglobin has not been accepted as explaining what occurs.

A theory of hydrogen sulphide action, advanced by Diakonow[29] and supported by Pohl[30] was that a reaction between hydrogen sulphide and the sodium bicarbonate of the blood plasma occurred whereby sodium sulphide was formed. They noted the similarity between the poisoning from hydrogen sulphide and that from sodium sulphide. Pohl believed that the sodium sulphide was carried in the blood. Haggard[31] in his studies definitely disproved this theory. He stated that "it appears that not only does hydrogen sulphide fail to form sodium sulphide when acting upon blood or plasma, but that a portion of the gas is actually destroyed." This is the form of an oxidation. Haggard believes that the products of oxidation combine, in part, with the sodium of the plasma. The oxygen is withdrawn from the corpuscles at such a rate that normally the hydrogen sulphide produced during digestion and absorption of sulphides, etc., is amply taken care of and poisoning does not result. In case of poisoning from hydrogen sulphide, however, "the greater the amount of inhaled hydrogen sulphide the more active will be the oxidation; but there will be also normally a higher concentration of hydrogen sulphide dissolved in the blood and in consequence a greater physiological effect." Haggard stated that the effect of poisoning is produced by the hydrogen sulphide held in solution in the blood and thus he corroborates the theory advanced by Laborde.

Kaufmann and Rosenthal,[32] in 1865, believed that the action of hydrogen sulphide was of such a nature as to result in oxygen hunger. They sought to demonstrate by an exhaustive experimental investigation that hydrogen sulphide poisoning is comparable to suffocation. It was pointed out, however, by Hoppe-Seyler[33] in a subsequent article that, while Kaufmann and Rosenthal defended this conception of suffocation, they did not account for the effect of hydrogen sulphide on the nervous system and, therefore, the explanation was not complete. Hoppe-Seyler believed that in warm-blooded animals the action of hydrogen sulphide on the oxyhemoglobin was very rapid. If the hydrogen sulphide was not in excess in case of poisoning in warm-blooded animals the effect was in the blood alone and there was no effect produced in the other tissues. Kaufman and Rosenthal pointed out that the action of hydrogen sulphide resembles suffocation very closely and the description given by Schäfer[34] of the symptoms of suffocation might readily be taken as a description of the symptoms of acute hydrogen sulphide poisoning.

In 1865 Eulenberg[35] subjected animals to toxic doses of hydrogen sulphide. He determined that 0.1 per cent of hydrogen sulphide was fatal for cats, rabbits, and doves within a short time. Young animals appeared to succumb to 0.05 per cent, and a dove was killed within four minutes by a concentration of 0.007 per cent; on the other hand, 0.014 per cent had no noticeable effect upon a young cat following 10 minutes' exposure, while 0.07 per cent asphyxiated a cat within 25 minutes and 0.11 per cent caused the death of another within 5 minutes. Eulenberg carefully reported the symptoms observed in cases of poisoning from different percentages of hydrogen sulphide and also recorded the pathological changes which he observed. He divided the poisoning into mild, medium, and severe, or asphyxia.

Biefel and Polek[36] some years later found that a rabbit died within 75 minutes when exposed to 0.05 per cent of hydrogen sulphide and concluded that 0.01 per cent was without effect. They observed crying, convulsions, trembling, respiratory disturbance, and an increase in the secretions of the salivary glands.

In 1884 Smirnow[37] reported that in his experiments he was unable to find the spectroscopic changes in the blood of animals poisoned by hydrogen sulphide, as reported by other investigators. He did not believe that the hemoglobin was in any way altered. The percentages of hydrogen sulphide used by him were reported as considerably higher than those given by the majority of the investigators. Smirnow stated that 0.3 per cent of hydrogen sulphide quickly kills small animals, while 0.2 per cent may cause death, and that 0.1 to 0.15 per cent may be endured for a considerable period. He studied the effect of hydrogen sulphide on tracheotomized animals; possibly this, together with poor methods for chemical analysis, might account for the higher percentages which he has reported.

The studies of Brouardel and Loye[38] were also performed on tracheotomized dogs, and they reported two types of death—one, fulminating, due apparently to direct action of the gas on the central nervous system, and the other, slow, with death due probably to asphyxia. They did not determine the absolute quantity of hydrogen sulphide breathed but depended upon the tension of the gas in air. They found 2 parts of hydrogen sulphide in 100 parts of air caused death within two to three minutes.

The attention of J. Peyron[39] was attracted to hydrogen sulphide poisoning because of the practice of injecting the gas into the rectum as a method of treatment of certain pulmonary diseases. He found that if a small amount of hydrogen sulphide gas was given no severe symptoms were produced and hydrogen sulphide could not be detected in the breath. However, if larger quantities were injected into the rectum a small part was liberated through the lungs, while the major part of the gas was fixed, presumably by the tissues. He believed that the appearance of the gas in the lungs depended upon its tension in the blood. Since with larger quantities of gas injected into the rectum, symptoms of poisoning developed, he concluded that rectal injections should be done with great care and only when absolutely necessary.

A. Flint[40] also studied the effect of hydrogen sulphide injected as an enemata. He reported that one-half fluid ounce of a saturated solution of hydrogen sulphide injected into the rectum of a dog was not a sufficient quantity to cause the gas to be eliminated through the lungs. If, however, larger quantities were injected the gas could be detected in the expired air. His results agreed with those of Peyron. In addition, Flint injected 1 fluid drachm of a saturated solution of hydrogen sulphide into the external jugular vein of a dog, whereupon hydrogen sulphide appeared in the breath. No objective symptoms of poisoning were noticed. He found that up to a certain limit hydrogen sulphide was destroyed in the blood in some unknown manner. He also observed that hydrogen sulphide had no inhibiting action upon the growth of bacteria and concluded that hydrogen sulphide would therefore have no destructive action on bacteria present in the lungs. Its use for the treatment of lung affections may therefore be considered as problematical.

Cahn[41] observed a striking case of hydrogen sulphide poisoning occurring in a student who carelessly exposed himself to the gas. The young man developed a severe abdominal pain which was followed by respiratory changes characteristic of hydrogen sulphide poisoning. Later, sugar appeared in the urine and persisted for three days, the young man finally recovering.

The symptoms produced by hydrogen sulphide poisoning would not be complete without mention of the mental depression which may occur. Wigglesworth[42] reported two cases of insanity caused by inhalation of hydrogen sulphide. They were characterized by great muscular excitement. One case recovered after five months, but the other had not recovered at the time of writing, although temporary improvement had been observed.

Perhaps the most exhaustive of all the experimental studies, with men as subjects, was made by Lehmann.[43] He subjected men to varying concentrations of hydrogen sulphide, ranging from 0.01 to 0.05 per cent, and observed severe poisoning. The symptoms reported were similar to those noted in animals exposed to hydrogen sulphide of the same percentages. He therefore concluded that the reaction of man to higher concentrations would be comparable to that of dogs subjected to like concentrations.

In 1908 Haibe[44] reported an interesting study of cases of chronic poisoning due to hydrogen sulphide occurring in the gas industry. These men presented symptoms of discomfort, depression, loss of appetite, pulmonary disturbances, gastric troubles, debility, and eventually icterus. Seven deaths were caused by hepatogenic icterus, while in those cases that recovered anemia was a constant finding. The men were apparently subjected to a relatively low concentration of the gas, although an analysis made in one location showed 0.063 per cent of hydrogen sulphide present. In addition he reported three cases all of which showed changes in the liver.

Sir Thomas Oliver[45] in 1911 investigated the sulphur mines of Sicily and reported a number of deaths—11—due to hydrogen sulphide poisoning. One boy was unconscious for several days and on recovery had lost his speech. Numerous cases of conjunctivitis occurred among the workmen at these mines.

In an experimental study on the effects of hydrogen sulphide upon animals (canary birds, white rats, guinea pigs, dogs, and goats) and upon men by Sayers, Mitchell, and Yant,[46] it was found that as low a concentration as 0.005 per cent would cause toxic symptoms and on continued exposure covering a number of days, with a concentration of 0.02 per cent, death occurred.

SUMMARY

The history of the study of hydrogen sulphide poisoning is of interest inasmuch as our present knowledge is built up from the work of many scientists. No one man may be credited with an epoch-making discovery, but each has laid a stone on which some other investigator has built. We now know that hydrogen sulphide is one of the most toxic of the gases. It is comparable to hydrogen cyanide in the rapidity of its action and the concentration from which death will result. In general, its action depends upon its concentration. In concentration of 0.005 it will cause poisoning. Hydrogen sulphide in such low percentage is often found in certain industries. It is, therefore, an industrial poison with which we should be well acquainted.

The exact mechanism of hydrogen sulphide poisoning is still unknown and is therefore a subject which invites further study. Such a study should be applied to those changes which occur in the body at the time poisoning occurs. Care should be exercised against inferring that a chemical change which may occur outside the living body may be comparable to the reactions occurring within the body. Such reasoning has been done in the past and has been discarded when careful experimental study on the living animal has proven the application to be incorrect.

BIBLIOGRAPHY ON HYDROGEN SULPHIDE POISONING

1. Rouelle, M. Sur l'air fixe et sur les effects dans certaines eaux minerales. Journal de Medicine, Vol. 39, pp. 449–464, 1773. Or Foureroy, Système de chimie.
2. Scheele, K. W. Ann. de Chim., Vol. XXV, p. 233, 1777.
3. Hallé, M. Recherches sur la nature du mephitisme des fosses d'aisance, 1785.
4. Dupuytren, M. Rapport sur une espéce de mephitisme des fosses d'aisance, produite par le gas azote. Journal de Medicine, Vol. XI, 1806, p. 187–213.
5. Chaussier, Franqois. Précis d'experiences faites sur les animaux avec le gaz hydrogène sulfuré. J. gen. de med., chir. et pharm. Paris, 1809, XV, 19–39.
6. Nysten, M. Cited by Thenard in Chimie, Vol. 4, 1827.
7. Dupuytren & Thenard. Thenard—Chimie, Vol. 4, 1827, p. 575. Also Dictionnaire des Sciences medicales, 1812, ii. 391.
8. Magendie, M. Cited by Thenard in his Chimie, Vol. 4.
9. Parent-Duchatelet. Rapport sur le curage des égouts Amelot, de la Roquette, Saint-Martin et autres. Annales d'Hygiene Publique, 1829, Series I, Vol. 2, pp. 1–159.
10. D'Arcet & Braconnot. Observations d'asphyxie lent due à l'insalubrité des habitations, et à des émanations metalliques. Ann. d'hyg., 1836, Vol. 16, 24–39.
11. Taylor, Alfred S. Sulphuretted hydrogen—Drains and Sewers, Thames Tunnel. Manual of Medical Jurisprudence, 1844, pp. 558–560; 1846, pp. 622–625. Also 1891.
12. Christison, Robert. A treatise on Poisons, in relation to medical jurisprudence, physiology, and the practice of physic. 2d Ed. XX, 1832. American edition, 1845, p. 617.
13. Bell, T. S. Case of Poisoning by Sulphuretted Hydrogen Gas. West's Journal of Medicine and Surgery, Louisville, 1851, 3s, VII, pp. 409–414.
14. Raphael, B. I. Two cases of poisoning by Hydrogen Sulphide gas. Transylv. Medical Journal, 1850–51, ii. 518–531.
15. Bernard, Claude. De l'elimination de l'hydrogène sulfuré par la surface pulmonarie. Archives générales de medecine, 5th Serie, Vol. 9, 1867, pp. 129–135.
16. Barker, T. Herbert. De l'influence des émanations des égouts. Extrait de la Sanitary Revue de Londres avril 1858 par le doctor Prosper de Pietra Santa. Annales d'Hygiene publique, 2d Serie, X, 1858, pp. 107–122.
17. Holden L. & Letheby, H. The medical history of the recent cases of poisoning in the Flectlane sewer. Lancet, Lond., 1861, 1, 187.
18. Hoppe-Seyler. Einwirkung des Schwefelwasserstoffgases auf das Blut. Centr. f. d. med. Wissensch., 1863, 433, No. 28.
19. Araki, T. Ueber den Blutfarbstoff and seine näheren Umwandlungsproducte. Ztschr. f. Physiol. Chem., 1889–90, 14, 405–416.
20. Gamgee, A. Schäfer, E. A., Textbook of physiology, 1898, 1, 249.
21. Lewisson. Zur Frage über Ozon im Blute. Virchow's Archiv, Bd. XXXVI, 1866, 15.
22. Kuhne. Lehrbuch der physiolog. Chemi. Leipzig, 1868, S. 215.
23. Laborde, J. V. Sur l'action physiologique et toxique de l'hydrogène sul. furé et en particulier sur le mecanisme de cette action. Compt. rend Soc. de biol., Par., 1886, 8, iii, pp. 113–116.

24. Harnack, E. (The action of hydrogen sulphide and acids on the coloring matter of the blood.) Ueber die Einwirkung des Schwefelwasserstoff und der Säuren auf den Blutfarbstoff. Zeit. f. physiol. Chem., 26, pp. 558–585, 1898–1899.
25. Clarke, T. W. & Hurtley, W. H. On Sulphemoglobin. J. Physiol. 1907, 36, 62.
26. Van Der Beigh. Enterogene Cyanose. Deut. Arch. f. klin. Med., 1905, 83, pp. 86–106.
27. West, S. & Clarke, W. Idiopathic cyanoses due to sulph-hemoglobinaemia. Med. Chirurg. Trans., 1907, Vol. 90, pp. 541–561.
28. Wallis, R. L. M. On Sulphaemoglobinaemia. Quart. J. Med., 1913–14, 7, 74.
29. Diakonow. (Relation of hydrogen sulphide to the organism. Med. Vestnik, St. Petersb., 1867). Trans. & abstr. by Hoppe-Seyler in Med.-Chem. Untersuch., 1866, 71, 251–254.
30. Pohl, Julius. Ueber die Wirkungsweise des Schwefelwasserstoffes und der Schwefelalkalien. Arch. Exp. Path. u. Pharmacol., 1886–87, 22, 1–25.
31. Haggard, H. W. The fate of sulfides in the blood. Jr. Biol. Chem., 1921, Vol. 49, p. 519.
32. Kaufmann, S., & Rosenthal, I. Ueber die Wirkungen des Schwefelwasserstoffgases auf den thierischen Organismus. Arch. f. Ant. Physiol. u. Wissensch. Med., Leipz., 1856, 659–675.
33. Hoppe-Seyler. Ueber die Einwirkung des Schwefelwasserstoffs auf den Blutfarbstoff. Med. Chem. Untersuch. a. d. Lab. zu Tübing., Berl., 1866, 151–159.
34. Schaefer, E. A. Text Book of Physiology. Asphyxia. 1898.
35. Eulenberg, H. Die Lehre von den schädlichen und giftigen Gasen. 1865, pp. 260–289.
36. Biefel, R., & Polek, TH. Ueber Kohlendunst und Leuchtgasvergiftung. Zeitschr. f. Biologie, 1880, Vol. 16, pp. 279–366.
37. Smirnow, L. Ueber die Wirkung des Schwefelwasserstoffes auf den thierischen Organismus. Centralb. f. die med. Wissensch., No. 37, p. 641, 1884.
38. Brouardel, P., & Loye, P. Recherches sur l'empoisonnement par l'hydrogène sulfuré. France Med., Par., 1885, ii, 1250–1253. Also Journal de Pharmac, et de Chimie, XII, p. 316, 1885.
39. Peyron, J. Du danger que peuvent presenter des injections d'hydrogène sulfuré. Compt. rend. Soc. biol., 1886, 3, 515–518.
40. Flint, A. On the elimination of sulphureted hydrogen artificially introduced into the body. Med. News, Phila., 1887, p. 670.
41. Cahn, A. Acute Schwefelwasserstoffvergiftung mit längerem Latenzstadium und sehr heftigen intestinalen Symptomen. Deutsches Archiv für klinische Medizin, 1883, XXXIV, p. 121.
42. Wigglesworth, J. Remarks on two cases of insanity caused by inhalation of hydrogen sulphide. Brit. M. J. Lond., 1892, ii, 124.
43. Lehmann, K. B. Experimentelle Studien über den Einfluss technisch und hygienisch wichitiger Gase und Dämpfe auf den Organismus. Theil V. Schwefelwasserstoff. Archiv für Hygiene, Vol. 14, 1892, pp. 135–189.
44. Haibe, A. Etude d'une series d'intoxications chroniques causées par le gaz sulfhydrique provenant de la production industriele du gaz pauvre. Academie Royale de Medicine de Belgique, Bulletin, Vol. 22, Bruxelles, 1908, pp. 535–544.

45. Oliver, Sir Thomas. The sulphur mines of Sicily: Their work, diseases and accident insurance. Brit. Med. Jour., July 1, 1911, Vol. II, p. 12.
46. Sayers, R. R., Mitchell. C. W., and Yant, W. P. Hydrogen sulphide as an industrial poison, Reports of Investigations, Serial No. 2491, June, 1923, Department of the Interior, Bureau of Mines.

ADDITIONAL REFERENCES

Akenfield, D. Centralbl. f. d. Wissensch., Berlin, 1885, No. 47.

Bell, T. S. Sulphureted hydrogen gas poisoning. West. J. M. & S., Louisville, 1851, 3. s., VIII, 19–36.

Bernard, Claude. Innocuité de l'hydrogène sulfuré introduit dans les voies digestives, cause de cette innoguité demontrie. Compt. rend. Societé de Biologie, 1856, iii, Serie ii, pp. 137–138.

Binet, P. Note sur la presence de la sulfo-methemoglobine dans l'empoisonnement par l'hydrogène sulfuré. Rev. med. de la Suisse Rom., Genève, 1896, XVI, 65–72.

Brown, Douglas. Petroleum Gas Poisoning. Medical Record, 1921, May 28, 99, p. 915.

Chantourelle, M. Reflexions sur l'action comparative des acides nitrique, du gas acide hydrosulfurique, etc. J. Gen. de Med., Chir. et Pharm., Par., 1819, LXVI, pp. 346–370.

Christison, Robert, & Turner, Edward. On the Effects of Poisonous Gases on Vegetables. Edin. Med. and Surg. Jour., 1827, XXVIII, pp. 356–364.

Guerard. Annales d'hygiène publique, 1840, XXIII, 131.

Harnack, E. Ueber Schwefelwasserstoffvergiftung. Arztl. Sachverständ. Z. 1897. 256.

Hermann. Lehrbuch der Toxikologie, 1874, S. 128.

Holtzmann. Die Möglichkeit der Schwefelwasserstoffvergiftung in Gerbereien. Zentralbl. f. Gewerbehyg., 1919, VII, 214.

Hoppe-Seyler. Beiträge zur Kentniss der indigobildenden Substanzen in Harn und des künstlichen Diabetes mellitus. Inaug.-Dissert., Berlin, 1883.

Hoppe-Seyler. Z. Physiol. Chemie, Berlin, 1881, p. 386.

Hoppe-Seyler & Thierfelder. Z. Physiol. Chem. Analyse 6, 1893, p. 283.

Husson, M. C. Compt. rend. Acad. d. Sc. Paris, T. LXXXI, p. 477.

Kobert, Rudolf. Pathologic-Anatomic demonstration of Intoxications on Corpses. Lehrbuch der Intoxikationen. Vol. I-II, pp. 97, 1902.

Kwilecki, A. Studium über die Giftigkeit des vom Menschen inhalierten Schwefelwasserstoffs mit besondere Rucksicht auf die Fabrikhygiene. Würzburg, 1890.

Laborde, J. V. De l'action physiologique et toxique des gaz dits mephetiques et en particulier du gaz hydrogène sulfuré. Tribune medicale, 1881, pp. 544–546, 591–594, 617–618.

Letheby, H. The fatal accident in the Fleet Lane sewer. Lancet, Lond., 1861, i, 455.

Lewin. Lehrbuch der Toxikologie, Wien, 1885.

Neumeister. Lehrbuch Physiol. Chemie, 2. Aufl., Jena, 1897, 578.

Nothnagel & Rossbach. Handbuch der Arzneimittellehre, 4. Aufl., 1880.

Orfila. Lehrbuch der Toxikologie, 5. Aufl., Traité des Exhumations, Vol. 1, p. 2, 1854.

Ozier. Questions relatives à la recherche de l'hydrogène sulfuré dans les Empoisonnements. Cong. internal med. lég., 1897, 2, 386–392.

Peyron, J. Variations que presente l'absorption de l'hydrogène sulfuré mis en contact de diverses surfaces chez l'animal vivant. Compt. rend. Soc. de Biol., 1885, ii, 556–558.

Peyron, J. De l'action toxique et physiologique de l'hydrogène sulfuré sur les animaux. Paris, 1888.

Peyron, J. De l'action de l'hydrogène sulfuré sur les mammiféres. Compt. rend. Soc. de biol., 1886, 3, 67–70, 515.

Prunelle. Extrait d'une observation sur le gaz hydrogène sufuré consideré comme cause de maladic. J. gen. de med., chir., et pharm., Par., 1800, XV, 39–42.

Saliiouski, E. Ueber die Entwickelung von Schwefelwasserstoff im Harn und das Verhalten des Schwefels in Organismus. Berl. klin. Wchnschr., 1888, 25, pp. 722–726.

Schultz, F. N. Zeit. f. physiol. Chem. 14, p. 449, 1898.

Skvortsoff, P. A. (Effect of hydrogen sulphide upon lung tissues in poisoning of animals by it.) St. Petersburg, 1896.

Smirnov, G. H. (Effect of hydrogen sulphide gas upon the human system, a supplement to the pathology of Cheyne-Stokes breathing.) Ehened. klin. gaz. St. Petersb., 1884, IV, 433–436.

Sonnenschein. Handbuch der gerichtl. Chemie. 2. Aufl., p. 295.

Stifft, H. Die physiologische und therapeutische Wirkung des Schwefelwasserstoffgases. Nach Beobachtungen an der kalten Schwefelguelle zu Weilback. Berl., 1886.

Surne, L. Asphyxie par l'hydrogène sulfuré dans un égout. Ann. d'hyg., Par., 1899, 3 s., XLI, pp. 253–257.

Thenard. By Guyton. Rapport sue le mémoire du cit. Thenart, concernant les differens états de l'oxide d'antimine, et ses combinaison avec l'hydrogène sulfuré. Annal. de Chimie, Vol. 32, p. 267.

Thompson. Occupational Diseases, 1914, p. 334.

Wilson. American Jour. of Pharmacy, Vol. LXV, 1893, No. 12, pp. 561–571.

References

1. Abd-el-Malek, Y., and S. G. Rizk. Bacterial sulphate reduction and the development of alkalinity. I. Experiments with synthetic media. J. Appl. Bacteriol. 26:7–13, 1963.
2. Abd-el-Malek, Y., and S. G. Rizk. Bacterial sulphate reduction and the development of alkalinity. II. Laboratory experiments with soils. J. Appl. Bacteriol. 26:14–19, 1963.
3. Adams, D. F., and R. K. Koppe. Direct GLC coulometric analysis of kraft mill gases. J. Air. Pollut. Control Assoc. 17:161–165, 1967.
4. Adams, D. F., F. A. Young, and R. A. Luhr. Evaluation of an odor perception threshold test facility. Tappi 51:62A–67A, 1968.
5. Adams, W. E. The Comparative Morphology of the Carotid Body and Carotid Sinus, p. 96. Springfield, Ill.: Charles C Thomas, 1958.
6. Adelman, I. R., and L. L. Smith, Jr. Effect of hydrogen sulfide on northern pike eggs and sac fry. Trans. Amer. Fish. Soc. 99:501–509, 1970.
7. Adelson, L., and I. Sunshine. Fatal hydrogen sulfide intoxication. Report of three cases occurring in a sewer. Arch. Path. 81:375–380, 1966.
8. Ahlborg, G. Hydrogen sulfide poisoning in shale oil industry. A.M.A. Arch. Ind. Hyg. Occup. Med. 3:247–266, 1951.
9. Alarie, Y. Sensory irritation of the upper airways by airborne chemicals. Toxicol. Appl. Pharmacol. 24:279–297, 1973.
10. Alexander, M. Microbial formation of environmental pollutants. Adv. Appl. Microbiol. 18:173, 1974.
11. Allam, A. I., and J. P. Hollis. Sulfide inhibition of oxidases in rice roots. Phytopathology 62:634–639, 1972.
12. American Conference of Governmental Industrial Hygienists (ACGIH). TLV® Threshold Limit Values for Chemical Substances and Physical Agents in the Workroom Environment with Intended Changes for 1975. Cincinnati: American Conference of Governmental Industrial Hygienists, 1975.
13. Andrews, J. C. Reduction of certain sulfur compounds to hydrogen sulfide by the intestinal microorganisms of the dog. J. Biol. Chem. 122:687–692, 1937–38.
14. Anichkov, S. V., and M. L. Belen'kii. (Translated by R. Crawford) Pharmacology of the Carotide Body Chemoreceptors, pp. 49–60. New York: The Macmillan Co., 1963.
15. Aves, C. M. Hydrogen sulphide poisoning in Texas. Texas State J. Med. 24:761–766, 1929.
16. Baba, I., K. Inada, and K. Tajima. Mineral nutrition and the occurrence of physiological diseases, pp. 173–195. In The Mineral Nutrition of the Rice Plant. Proceedings of a Symposium at the International Rice Research Institute, February 1964. Baltimore: Johns Hopkins Press, 1965.
17. Bamesberger, W. L., and D. F. Adams. Improvements in the collection of hydrogen sulfide in cadmium hydroxide suspension. Environ. Sci. Technol. 3:258–261, 1969.

18. Barthelemy, H. L. Ten years' experience with industrial hygiene in connection with the manufacture of viscose rayon. J. Ind. Hyg. Toxicol. 21:141–151, 1939.
19. Barton, L. V. Toxicity of ammonia, chlorine, hydrogen cyanide, hydrogen sulphide, and sulphur dioxide gases. IV. Seeds. Contrib. Boyce Thompson Inst. 11:357–363, 1940.
20. Basch, F. Über Schwefelwasserstoffvergiftung bei äusserlicher Applikation von elementarem Schwefel in Salbenform. Naunyn-Schmiedebergs Arch. Exp. Pathol. Pharmakol. 111:126–132, 1926.
21. Baxter, C. F., and R. van Reen. Some aspects of sulfide oxidation by rat-liver preparations. Biochim. Biophys. Acta 28:567–573, 1958.
22. Baxter, C. F., and R. van Reen. The oxidation of sulfide to thiosulfate by metallo-protein complexes and by ferritin. Biochim. Biophys. Acta 28:573–578, 1958.
23. Baxter, C. F., R. Van Reen, P. B. Pearson, and C. Rosenberg. Sulfide oxidation in rat tissues. Biochim. Biophys. Acta 27:584–591, 1958.
24. Beasley, R. W. R. The eye and hydrogen sulphide. Brit. J. Ind. Med. 20:32–34, 1963.
25. Beerman, H. Some physiological actions of hydrogen sulphide. J. Exp. Zool. 41:33–43, 1924.
26. Benedict, H. M., and W. H. Breen. The use of weeds as a means of evaluating vegetation damage caused by air pollution, pp. 177–190. In Proceedings of the Third National Air Pollution Symposium. Sponsored by Stanford Research Institute, Pasadena, California, April 1955.
27. Berglund, B. Quantitative and qualitative analysis of industrial odors with human observers. Ann. N.Y. Acad. Sci. 237:35–51, 1974.
28. Berglund, B., U. Berglund, G. Ekman, and T. Engen. Individual psychophysical functions for 28 odorants. Percept. Psychophys. 9:379–384, 1971.
29. Berglund, B., U. Berglund, T. Engen, and T. Lindvall. The effect of adaptation on odor detection. Percept. Psychophys. 9:435–438, 1971.
30. Berglund, B., U. Berglund, E. Jonsson, and T. Lindvall. On the Scaling of Annoyance to Environmental Factors. Department of Psychology Number 451. Stockholm, Sweden: University of Stockholm, 1975. 10 pp.
31. Berglund, B., U. Berglund, and T. Lindvall. Perceptual interaction of odors from a pulp mill, pp. A40–A43. In Proceedings of the 3rd International Clean Air Congress. Düsseldorf: VDI-Verlag, 1973.
32. Berglund, B., U. Berglund, and T. Lindvall. A psychological detection method in environmental research. Environ. Res. 7:342–352, 1974.
33. Berglund, B., U. Berglund, and T. Lindvall. Measurement of rapid changes of odor concentration by a signal detection approach. J. Air Pollut. Control Assoc. 24:162–164, 1974.
34. Berglund, B., U. Berglund, T. Lindvall, and L. T. Svensson. A quantitative principle of perceived intensity summation in odor mixtures. J. Exp. Psychol. 100:29–38, 1973.
35. Bergstermann, H., and H. D. Lummer. Die Wirkung von Schwefelwasserstoff und seinen Oxydationsprodukten auf Bernsteinsäuredehydrase. Arch. Exp. Path. Pharmakol. 204:509–519, 1947.
36. Birkinshaw, J. H., W. P. K. Findlay, and R. A. Webb. Biochemistry of the wood-rotting fungi. 3. The production of methyl mercaptan by *Schizophyllum commune* Fr. Biochem. J. 36:526–529, 1942.

37. Bittersohl, G. Beitrag zum toxischen Wirkungsmechanismus von Schwefelwasserstoff. Z. Gesamte Hyg. Ihre Grenzgeb. 17:305–308, 1971.
38. Blanchette, A. R., and A. D. Cooper. Determination of hydrogen sulfide and methyl mercaptan in mouth air at the parts-per-billion level by gas chromatography. Anal. Chem. 48:729–731, 1976.
39. Bock, R., and H-J. Puff. Bestimmung von Sulfid mit einer sulfidionenempfindlichen Elektrode. Fresenius Z. Anal. Chem. 240:381–386, 1968.
40. Bonn, E. W., and B. J. Follis. Effects of hydrogen sulfide on channel catfish, *Ictalurus punctatus*. Trans. Amer. Fish. Soc. 96:31–36, 1967.
41. Boström, C-E. The absorption of low concentrations (pphm) of hydrogen sulfide in a $Cd(OH)_2$-suspension as studied by an isotopic tracer method. Air Water Pollut. Int. J. 10:435–441, 1966.
42. Boyland, E., and E. Gallico. Catalase poisons in relation to changes in radiosensitivity. Brit. J. Cancer 6:160–172, 1952.
43. Breysse, P. A. Hydrogen sulfide fatality in a poultry feather fertilizer plant. Amer. Ind. Hyg. Assoc. J. 22:220–222, 1961.
44. British Sulphur Corporation Limited. World Survey of Sulphur Resources. London: British Sulphur Corporation Limited, 1966. 148 pp.
45. Brock, T. D., K. M. Brock, R. T. Belly, and R. L. Weiss. *Sulfolobus*: A new genus of sulfur-oxidizing bacteria living at low pH and high temperature. Arch. Mikrobiol. 84:54–68, 1972.
46. Brodie, D. A., and H. L. Borison. Analysis of central control of respiration by the use of cyanide. J. Pharmacol. Exp. Ther. 118:220–229, 1956.
47. Brody, S. S., and J. E. Chaney. Flame photometric detector. The application of a specific detector for phosphorus and for sulfur compounds sensitive to subnanogram quantities. J. Gas Chromatogr. 4:2:42–46, 1966.
48. Brunold, C., and K. H. Erismann. H_2S als Schwefelquelle bei *Lemna minor* L: Einfluss auf das Wachstum, den Schwefelgehalt und Sulfataufnahme. Experientia 30:465–467, 1974.
49. Brunold, C., and K. H. Erismann. H_2S as sulfur source in *Lemna minor* L.: II. Direct incorporation into cysteine and inhibition of sulfate assimilation. Experientia 31:508–510, 1975.
50. Buchanan, R. E., and N. E. Gibbons, Eds. Bergey's Manual of Determinative Bacteriology. (8th ed.) Baltimore: Williams & Wilkins Co. 1974. 1246 pp.
51. Buck, M., and H. Gies. Die Messung von Schwefelwasserstoff in der Atmosphäre—Kombinierte H_2S- und SO_2-Messung. Staub 26(9):379–384, 1966.
52. Buck, M., and H. Stratmann. Bestimmung von Schwefelwasserstoff in der Atmosphäre. Staub 24(7):241–250, 1964.
53. Butlin, K. R. The bacterial sulphur cycle. Res. Sci. Appl. Ind. 6:184–191, 1953.
54. Butlin, K. R., and J. R. Postgate. The microbiological formation of sulphur in Cyrenaican lakes, pp. 112–122. In J. L. Cloudsley-Thompson, Ed. Biology of Deserts. Proceedings of a Symposium on the Biology of Hot and Cold Deserts. London: Institute of Biology, 1954.
55. Cabanac, M. Physiological role of pleasure. Science 173:1103–1107, 1971.
56. Cain, W. S. Odor intensity after self-adaptation and cross-adaptation. Percept. Psychophys. 7:271–275, 1970.
57. Cain, W. S. Contribution of the trigeminal nerve to perceived odor magnitude. Ann. N.Y. Acad. Sci. 237:28–34, 1974.

58. Cain. W. S. Odor intensity: Mixtures and masking. Chem. Senses Flavor 1:339–352, 1975.
59. Cain, W. S., and M. Drexler. Scope and evaluation of odor counteraction and masking. Ann. N.Y. Acad. Sci. 237:427–439, 1974.
60. Campbell, C. L., R. K. Dawes, S. Deolalkar, and M. C. Merritt. Effect of certain chemicals in water on the flavor of brewed coffee. Food Res. 23:575–579, 1958.
61. Cantor, M. O., and J. E. Weiler. Effect of hydrogen sulfide upon intestinal decompression tubes. An *in vivo* and *in vitro* study. Amer. J. Gastroenterol. 38:583–586, 1962.
62. Carson, M. B. Hydrogen sulfide exposure in the gas industry. Ind. Med. Surg. 32:63–64, 1963.
63. Cederlöf, R., M.-L. Edfors, L. Friberg, and T. Lindvall. Determination of odor thresholds for flue gases from a Swedish sulfate cellulose plant. Tappi 48:405–411, 1965.
64. Cederlöf, R., L. Friberg, E. Jonsson, L. Kaij, and T. Lindvall. Studies of annoyance connected with offensive smell from a sulphate cellulose factory. Nord. Hyg. Tidskr. 45:39–48, 1964.
65. Cederlöf, R., E. Jonsson, and S. Sörensen. On the influence of attitudes to the source on annoyance reactions to noise. A field experiment. Nord. Hyg. Tidskr. 48:46–59, 1967.
66. Chance, B., and B. Schoener. High and low energy states of cytochromes. I. In mitochondria. J. Biol. Chem. 241:4567–4573, 1966.
67. Chiarenzelli, R. V., and E. L. Joba. The effects of air pollution on electrical contact materials: A field study. J. Air Pollut. Control Assoc. 16:123–127, 1966.
68. Clarke, T. W., and W. H. Hurtley. On sulphaemoglobin. J. Physiol. 36:62–67, 1907.
69. Cohen, G., and P. Hochstein. Glutathione peroxidase: Inverse temperature dependence and inhibition by sulfide and penicillamine. Fed. Proc. 24:605, 1965. (abstr.)
70. Cohen, Y., and H. Delassue. Etude comparative du métabolisme du ^{35}S chez la Souris après administration par voie orale ou sous-cutanée de radiosulfate et de radiosulfure de sodium. C. R. Soc. Biol. 153:999–1003, 1959.
71. Colby, P. J., and L. L. Smith, Jr. Survival of walleye eggs and fry on paper fiber sludge deposits in Rainy River, Minnesota. Trans. Amer. Fish. Soc. 96:278–296, 1967.
72. Cooper, P. Carbon disulphide toxicology: The present picture. Food Cosmet. Toxicol. 14:57–59, 1976.
73. Cooper, R. C. Photosynthetic bacteria in waste treatment. Dev. Ind. Microbiol. 4:95–103, 1963.
74. Cooper, R. C., D. Jenkins, and L. Young. Aquatic Microbiology Laboratory Manual. Austin: Association of Environmental Engineering Professors, University of Texas, 1976. [200 pp.]
75. Corbit, T. E., and T. Engen. Facilitation of olfactory detection. Percept. Psychophys. 10:433–436, 1971.
76. Cordon, T. C. Leather and fur processing, pp. 505–510. In McGraw-Hill Encyclopedia of Science and Technology. Vol. 7. (4th ed.) New York: McGraw-Hill, Inc., 1977.
77. Coryell, C. D. The existence of chemical interactions between the hemes in

ferrihemoglobin (methemoglobin) and the role of interactions in the interpretation of ferro-ferrihemoglobin electrode potential measurements. J. Phys. Chem. 43:841–852, 1939.

78. Coryell, C. D., F. Stitt, and L. Pauling. The magnetic properties and structure of ferrihemoglobin (methemoglobin) and some of its compounds. J. Amer. Chem. Soc. 59:633–642, 1937.
79. Crider, W. L. Hydrogen flame emission spectrophotometry in monitoring air for sulfur dioxide and sulfuric acid aerosol. Anal. Chem. 37:1770–1773, 1965.
80. Czurda, V. Schwefelwasserstoff als ökolgischer Faktor der Algen. Zentralbl. Bakteriol. Parasitenk. Infektionskr. 103:285–311, 1941.
81. Danhof, I. E. The clinical gas syndromes: A pathophysiologic approach. Ann. N.Y. Acad. Sci. 150:127–140, 1968.
82. Deaths at a rendering plant—Ohio. Morbid. Mortal. Week. Rep. 24:435–436, 1975.
83. de Cormis, L., and J. Bonte. Etude du degagement d'hydrogène sulfuré par des feuilles de plantes ayant reçu du dioxyde de soufre. C. R. Acad. Sci. 270D:2078–2080, 1970.
84. Delwiche, E. A. A micromethod for the determination of hydrogen sulfide. Anal. Biochem. 1:397–401, 1960.
85. Denis, W., and L. Reed. The action of blood on sulfides. J. Biol. Chem. 72:385–394, 1927.
86. Dobrovolsky, I. A., and E. A. Strikha. Study of phytotoxicity of some components of industrial pollution of air. Ukr. Bot. Zh. 27:640–644, 1970. (in Russian, summary in English)
87. Dorland's Illustrated Medical Dictionary. (25th ed.) Philadelphia: W. B. Saunders Company, 1974. 1748 pp.
88. Drabkin, D. L., and J. H. Austin. Spectrophotometric studies. II. Preparations from washed blood cells; nitric oxide hemoglobin and sulfhemoglobin. J. Biol. Chem. 112:51–65, 1935–1936.
89. Dubois, L., and J. L. Monkman. The analysis of airborne pollutants. Background Paper D25-3. In Background papers prepared for the national conference on Pollution and our Environment held in Montreal from Oct. 31 to Nov. 4, 1966. Toronto: Canadian Council of Resource and Environment Ministers, 1966.
90. Dynamic calibration of air analysis systems, pp. 20–31. In Methods of Air Sampling and Analysis. Washington, D.C.: American Public Health Association, 1972.
91. Dziewiatkowski, D. D. Fate of ingested sulfide sulfur labeled with radioactive sulfur in the rat. J. Biol. Chem. 161:723–729, 1945.
92. Ekman, G., B. Berglund, U. Berglund, and T. Lindvall. Perceived intensity of odor as a function of time of adaptation. Scand. J. Psychol. 8:177–186, 1967.
93. Engen, T. Psychophysics. I. Discrimination and detection, pp. 11–46. In J. W. Kling and L. A. Riggs, Eds. Woodworth and Schlosberg's Experimental Psychology. (3rd ed.) New York: Holt, Rinehart and Winston, Inc. 1971.
94. Engen, T. Use of sense of smell in determining environmental quality, pp. 133–146. In W. A. Thomas, Ed. Indicators of Environmental Quality. New York: Plenum Press, 1972.
95. Engen, T. The sense of smell. Annu. Rev. Psychol. 24:187–206, 1973.

96. Engen, T. Taste and smell, pp. 554–561. In J. E. Birren and K. Warner Schaie, Eds. Handbook of the Psychology of Aging. New York: Van Nostrand Reinhold Co., 1977.
97. Engen, T., and T. N. Bosack. Facilitation in olfactory detection. J. Comp. Physiol. Psychol. 68:320–326, 1969.
98. Engen, T., and L. P. Lipsitt. Decrement and recovery of responses to olfactory stimuli in the human neonate. J. Comp. Physiol. Psychol. 59:312–316, 1965.
99. Engen, T., and D. H. McBurney. Magnitude and category scales of the pleasantness of odors. J. Exp. Psychol. 68:435–440, 1964.
100. Engen, T., and B. M. Ross. Long-term memory of odors with and without verbal descriptions. J. Exp. Psychol. 100:221–227, 1973.
101. Epple, G. Olfactory communication in South American primates. Ann. N.Y. Acad. Sci. 237:261–278, 1974.
102. Espach, R. H. Sources of hydrogen sulfide in Wyoming. Ind. Eng. Chem. 42:2235–2237, 1950.
103. Evans, C. L. The toxicity of hydrogen sulphide and other sulphides. Q. J. Exp. Physiol. 52:231–248, 1967.
104. Evelyn, K. A., and H. T. Malloy. Microdetermination of oxyhemoglobin, methemoglobin, and sulfhemoglobin in a single sample of blood. J. Biol. Chem. 126:655–662, 1938.
105. Fairchild, E. J., II, S. D. Murphy, and H. E. Stokinger. Protection by sulfur compounds against the air pollutants ozone and nitrogen dioxide. Science 130:861–862, 1959.
106. Falgout, D. A., and C. I. Harding. Determination of H_2S exposure by dynamic sampling with metallic silver filters. J. Air Pollut. Control Assoc. 18:15–20, 1968.
107. Faller, N. Schwefeldioxid, Schwefelwasserstoff, nitrose Gase und Ammoniak als ausschliessliche $S^{=}$ bzw. N-Quellen der höheren Pflanze. Z. Planzenernähr. Dueng. Bodenk. 131:120–130, 1972.
108. Fankhauser, H., C, Brunold, and K. H. Erismann. Subcellular localization of *o*-acetylserine sulfhydrylase in spinach leaves. Experientia 32:1494–1497, 1976.
109. Fechner, G. T. Elemente der Psychophysik. Leipzig: Druck and Verlag von Breitkopf und Härtel, 1860. 571 pp.
110. Finch, C. A. Methemoglobinemia and sulfhemoglobinemia. N. Engl. J. Med. 239:470–478, 1948.
111. Ford, H. W. Levels of hydrogen sulfide toxic to citrus roots. J. Amer. Soc. Hort. Sci. 98:66–68, 1973.
112. Frendo, J. The role of elementary sulphur and erythrocyte glutathione in the formation of sulphaemoglobin. Clin. Chim. Acta 24:1–4, 1969.
113. Friberg, L., E. Jonsson, and R. Cederlöf. Studies of hygienic nuisances of waste gases from sulfate plup mill. Part I. An interview investigation. Nord Hyg. Tidskr. 41(3–4):41–50, 1960. (in Swedish)
114. Friberg, L., E. Jonsson, and R. Cederlöf. Studies of hygienic nuisances of waste gases from sulfate pulp mill. Part II. Odor threshold determinations for waste gases. Nord. Hyg. Tidskr. 41(3–4):50–62, 1960. (in Swedish)
115. Fyn-Djui, D. Basic data for the determination of limit of allowable concentration of hydrogen sulfide in atmosphere, pp. 66–73. In B. S. Levine [Translator], U.S.S.R. Literature on Air Pollution and Related Occupational Diseases. A Survey. Vol. 5. Washington, D.C.: U.S. Public Health Service, 1961.

116. Gas chromatography, pp. 99–109. In Methods of Air Sampling and Analysis. Washington D.C.: American Public Health Association, 1972.
117. Gassman, M. L. A reversible conversion of phototransformable protochlorophyll(ide)$_{650}$ to photoinactive protochlorophyll(ide)$_{633}$ by hydrogen sulfide in etiolated bean leaves. Plant Physiol. 51:139–145, 1973.
118. Gilardi, E. F., and R. M. Manganelli. A laboratory study of a lead acetate-tile method for the quantitative measurement of low concentrations of hydrogen sulfide. J. Air Pollut. Control Assoc. 13:305–309, 1963.
119. Giovanelli, J., and S. H. Mudd. Sulfuration of O-acetylhomoserine and O-acetylserine by two enzyme fractions from spinach. Biochem. Biophys. Res. Commun. 31:275–287, 1968.
120. Gloor, P. Temporal lobe epilepsy: Its possible contribution to the understanding of the functional significance of the amygdala and its interaction with neocortical-temporal mechanisms, pp. 423–457. In B. E. Eleftheriou, Ed. The Neurobiology of the Amygdala. Proceedings of a Symposium, 1971. New York: Plenum Press, 1972.
121. Goldsack, D. E., W. S. Eberlein, and R. A. Alberty. Temperature jump studies of sperm whale metmyoglobin. III. Effect of heme-linked groups of ligand binding. J. Biol. Chem. 241:2653–2660, 1966.
122. Goren, S. Plants pollute air. J. Air Pollut. Control Assoc. 9:105–109, 1959.
123. Gosselin, R. E., H. C. Hodge, R. P. Smith, and M. N. Gleason. Clinical Toxicology of Commercial Products (4th ed.) Baltimore: The Williams & Wilkins Co., 1976. [1783 pp.]
124. Grennefelt, P., and T. Lindvall. Sensory and physical-chemical studies of pulp mill odors, pp. A36–A39. In Proceedings of the 3rd International Clean Air Congress. Düsseldorf: VDI-Verlag, 1973.
125. Grennfelt, P., and T. Lindvall. A sensory and physical-chemical survey of odorous effluents from a kraft pulp mill. Sven. Papperstidn. 15:563–569, 1974.
126. Gunter, A. P. The therapy of acute hydrogen sulfide poisoning. Chem. Abstr. 50:5916f, 1956.
127. Haggard, H. W. The fate of sulfides in the blood. J. Biol. Chem. 49:519–529, 1921.
128. Haggard, H. W. The toxicology of hydrogen sulphide. J. Ind. Hyg. 7:113–121, 1925.
129. Haggard, H. W., Y. Henderson, and T. J. Charlton. The influence of hydrogen sulphide upon respiration. Amer. J. Physiol. 61:289–297, 1922.
130. Hamauzu, Y. Odor perception measurement by the use of odorless room. Sangyo Kagai (Ind. Publ Nuisance) 5:718–723, 1969. (in Japanese)
131. Hartmann, C. H. Improved chromatographic techniques for sulfur pollutants. Paper 71-1046 Presented at the Joint Conference on Sensing of Environmental Pollutants, Palo Alto, California, Nov. 8–10, 1971. 6 pp.
132. Hayes, W. J., Jr. Tests for detecting and measuring long-term toxicity, pp. 65–77. In W. J. Hayes, Jr., Ed. Essays in Toxicology. Vol. 3. New York: Academic Press, 1972.
133. Henderson, Y., and H. W. Haggard. Noxious Gases and the Principles of Respiration Influencing their Action. (2nd ed.) New York: Reinhold Publishing Corporation, 1943. 294 pp.
134. Hendrickson, H. R., and E. E. Conn. Cyanide metabolism in higher plants. IV. Purification and properties of the β-cyanoalanine synthase of blue lupine. J. Biol. Chem. 244:2632–2640, 1969.

135. Henion, K. E. Odor pleasantness and intensity: A single dimension. J. Exp. Psychol. 90:275–279, 1971.
136. Heymans, C., J-J. Bouckaert, and L. Dautrebande. Au sujet du mécanisme de la stimulation respiratoire par le sulfure de sodium. C. R. Soc. Biol. 106:52–54, 1931.
137. Heymans, C., J. J. Bouckaert, U.S. v. Euler, and L. Dautrebande. Sinus carotidiens et réflexes vasomoteurs. Arch. Int. Pharmacodyn. Ther. 43:86–110, 1932.
138. Heymans, C., and E. Neil. Cardiovascular reflexes of chemoreceptor origin, pp. 176–184. In Reflexogenic Areas of the Cardiovascular System. London: J. & A. Churchill Ltd., 1958.
139. High, M. D., and S. W. Horstman. Field experience in measuring hydrogen sulfide. Amer. Ind. Hyg. Assoc. J. 26:366–373, 1965.
140. Hill, A. C. Vegetation: A sink for atmospheric pollutants. J. Air Pollut. Control Assoc. 21:341–346, 1971.
141. Horstman, S. W., R. F. Wromble, and A. N. Heller. Identification of community odor problems by use of an observer corps. J. Air Pollut. Control Assoc. 15:261–264, 1965.
142. Hugo, V. [Description of sewers in section Jean Valjean], pp. 199, 250. In Les Miserables. Vol. IV. [Translated from French] Boston: Little, Brown and Company, 1887.
143. Huovinen, J. A., and B. E. Gustafsson. Inorganic sulphate, sulphite and sulphide as sulphur donors in the biosynthesis of sulphur amino acids in germ-free and conventional rats. Biochim. Biophys. Acta 136:441–447, 1967.
144. Hurwitz, L. J., and G. I. Taylor. Poisoning by sewer gas with unusual sequelae. Lancet 1:1110–1112, 1954.
145. Illinois Institute for Environmental Quality. Hydrogen Sulfide Health Effects and Recommended Air Quality Standard. Document No. 74-24. Chicago: State of Illinois, Institute for Environmental Quality, 1974. 27 pp.
146. Intersociety Committee. Methods of ambient air sampling and analysis. Tentative method of gas chromatographic analysis for sulfur-containing gases in the atmosphere (automatic method with flame photometer detector). Health Lab. Sci. 10:241–250, 1973.
147. Intersociety Committee. Methods of ambient air sampling and analysis. Tentative method of analysis for sulfur-containing gases in the atmosphere (automatic method with flame photometer detector). Health Lab. Sci. 10:342–348, 1973.
148. Jacobs, M. B. Techniques for measurement of hydrogen sulfide and sulfur oxides, pp. 24–36. In J. P. Lodge, Jr., Ed. Atmospheric Chemistry of Chlorine and Sulfur Compounds. Proceedings of a Symposium held at Robert A. Taft Engineering Center, Cincinnati, Ohio, Nov. 4–6, 1957. Geophysical Monograph Number 3. Washington D.C.: American Geophysical Union, 1959.
149. Jacobs, M. B. The Analytical Toxicology of Industrial Inorganic Poisons, p. 543. New York: Interscience Publishers, 1967.
150. Jacobs, M. B., M. M. Braverman, and S. Hochheiser. Ultramicrodetermination of sulfides in air. Anal. Chem. 29:1349–1351, 1957.
151. Jacques, A. G. The kinetics of penetration. XII. Hydrogen sulfide. J. Gen. Physiol. 19:397–418, 1936.

152. Jandl, J. H., L. K. Engle, and D. W. Allen. Oxidative hemolysis and precipitation of hemoglobin. I. Heinz body anemias as an acceleration of red cell aging. J. Clin. Invest. 39:1818–1836, 1960.
153. Johnson, E. A. The reversion of haemoglobin of sulphaemoglobin and its coordination derivatives. Biochim. Biophys. Acta 207:30–40, 1970.
154. Jones, F. N. Olfactory absolute thresholds and their implication for the nature of the receptor process. J. Psychol. 40:223–227, 1955.
155. Jones, F. N., and M. H. Woskow. On the intensity of odor mixtures. Ann. N.Y. Acad. Sci. 116:484–494, 1964.
156. Jope, E. M. The disappearance of sulphemoglobin from blood of TNT workers in relation to dynamics of red cell destruction. Brit. J. Ind. Med. 3:136–142, 1946.
157. Joshi, M. M., and J. P. Hollis. Interaction of *Beggiatoa* and rice plant: Detoxification of hydrogen sulfide in the rice rhizosphere. Science 195:179–180, 1977.
158. Joshi, M. M., I. K. A. Ibrahim, and J. P. Hollis. Hydrogen sulfide: Effects on the physiology of rice plants and relation to straighthead disease. Phytopathology 65:1165–1170, 1975.
159. Kadota, H., and Y. Ishida. Production of volatile sulfur compounds 1591 by microorganisms. Annu. Rev. Microbiol. 26:127–138, 1972.
160. Kaipainen, W. J. Hydrogen sulfide intoxication. Rapidly transient changes in the electrocardiogram suggestive of myocardial infarction. Ind. Hyg. Dig. 19:Abstr. 529, 1955.
161. Kaiser, E. R. Odor and its measurement, pp. 509–527. In A. C. Stern, Ed. Air Pollution. Vol. 1. New York: Academic Press, 1962.
162. Katz, M. Analysis of inorganic gaseous pollutants, p. 78. In A. C. Stern, Ed. Air Pollution. Vol. II. (2nd ed.) New York: Academic Press, 1968.
163. Katz. M. Hydrogen sulfide, pp. 81–82. In Measurement of Air Pollutants: Guide to the Selection of Methods. Geneva: World Health Organization, 1969.
164. Keilin, D. Cytochrome and respiratory enzymes. Proc. Roy. Soc. London B104:206–252, 1928.
165. Keilin, D. On the combination of methaemoglobin with H_2S. Proc. Roy. Soc. London B113:393–404, 1933.
166. Keilin, D., and E. F. Hartree. Uricase, amino acid oxidase, and xanthine oxidase. Proc. Roy. Soc. London B119:114–140, 1936.
167. Kellogg, W. W., R. D. Cadle, E. R. Allen, A. L. Lazrus, and E. A. Martell. The sulfur cycle. Science 175:587–596, 1972.
168. Kemper, F. D. A near-fatal case of hydrogen sulfide poisoning. Can. Med. Assoc. J. 94:1130–1131, 1966.
169. Kirshenbaum, I. Heavy water, pp. 432–434. In McGraw-Hill Encyclopedia of Science and Technology. Vol. 6. (4th ed.) New York: McGraw-Hill, Inc., 1977.
170. Kleinfeld, M., C. Giel, and A. Rosso. Acute hydrogen sulfide intoxication; an unusual source of exposure. Ind. Med. Surg. 33:656–660, 1964.
171. Kohgo, T., R. Endo, T. Oyake, and H. Shiradawa. Research on the odor nuisance in Hokkaido. Part 1. The effect of the odor on the environment and residence. Taiki Osen Kenkyu (J. Jap. Soc. Air Pollut.) 2:51, 1967. (in Japanese)
172. Kolthoff, I. M., and P. J. Elving, Eds. Treatise on Analytical Chemistry. Part II. Analytical Chemistry of the Elements. Vol. 7. Sulfur. Selenium.

Tellurium. Fluorine. The Halogens. Manganese. Rhenium. New York: Interscience Publishers, 1961. 567 pp.

173. Koppanyi, T., and C. R. Linegar. Contribution to the pharmacology of sulfides. Fed. Proc. 1:155–156, 1942. (abstr.)

174. Koppe, R. K., and D. F. Adams. Evaluation of gas chromatographic columns for analysis of subparts per million concentrations of gaseous sulfur compounds. Environ. Sci. Technol. 1:479–481, 1967.

175. Korochanskaya, S. P. Hydrogen sulfide oxidation with the blood and tissues. Farmakol. Toxshikol. 28:490–492, 1965. (in Russian, summary in English)

176. Kośmider, S., E. Rogala, and A. Pachołek. Electrocardiographic and histochemical studies of the heart muscle in acute experimental hydrogen sulfide poisoning. Arch. Immunol. Ther. Exp. (Warsz) 15:731–740, 1967.

177. Köster, E. P. Adaptation and Cross-Adaptation in Olfaction. An Experimental Study with Olfactory Stimuli at Low Levels of Intensity. Ph.D. Thesis. Utrecht, Netherlands: University of Utrecht, 1971. 212 pp.

178. Kruszyna, H., R. Kruszyna, and R. P. Smith. Calibration of a turbidimetric assay for sulfide. Anal. Biochem. 69:643–645, 1975.

179. Kuznetsov, S. I., M. V. Ivanov, and N. N. Lyalikova. (C. H. Oppenheimer, Editor of English edition, P. T. Broneer, Translator) Introduction to Geological Microbiology. New York: McGraw-Hill Book Company, Inc., 1963. 251 pp.

180. Laffort, P., and A. Dravnieks. An approach to a physico-chemical model of olfactory stimulation in vertebrates by single compounds. J. Theor. Biol. 38:335–345, 1973.

181. Lahmann, E. Methoden der Messung gasförmiger Luftverunreinigungen. Staub 25(9):346–351, 1965.

182. Larsen, V. Une endémie d'affections oculaires provoquées par l'hydrogène sulfuré chez des ouvriers travaillant à un tunnel. Acta Ophthalmol. 41:271–286, 1943–1944.

183. Larson, C. P., C. C. Reberger, and M. J. Wicks. The purple brain death. Med. Sci. Law 4:200–202, 1964.

184. Laug, E. P., and J. H. Draize. The percutaneous absorption of ammonium hydrogen sulfide and hydrogen sulfide. J. Pharmacol. Exp. Ther. 76:179–188, 1942.

185. Leathen, W. W., N. A. Kinsel, and S. A. Braley. *Ferrobacillus ferrooxidans*: A chemosynthetic autotrophic bacterium. J. Bacteriol. 72:700–704, 1956.

186. Lehmann, K. B. Experimentelle Studien über den Einfluss technisch und hygienisch wichtiger Gase und Dämpfe auf den Organismus. Theil V. Schwefelwasserstoff. Arch. Hyg. 14:135–189, 1892.

187. Leithe, W. The Analysis of Air Pollutants, p. 143. Ann Arbor: Humphrey Science Publishers, Inc., 1970.

188. Lendle, L. Wirkungsbedingungen von Blausäure und Schwefelwasserstoff und Möglichkeiten der Vergiftungsbehandlung. Jap. J. Pharmacol. 14:215–224, 1964.

189. Levine, S. Nonperipheral chemoreceptor stimulation of ventilation by cyanide. J. Appl. Physiol. 39:199–204, 1975.

190. Levitt, M. D., R. B. Lasser, J. S. Schwartz, and J. H. Bond. Studies of a flatulent patient. N. Engl. J. Med. 295:260–262, 1976.

191. Lindvall, T. Measurement of odorous air pollutants. Nord. Hyg. Tidskr. 47:41–71, 1966. (in Swedish)

192. Lindvall, T. Nuisance effects of air pollutants. Nord. Hyg. Tidskr. 50(3):99–114, 1969. (in Swedish)

193. Lindvall, T. On sensory evaluation of odorous air pollutant intensities. Measurements of odor intensity in the laboratory and in the field with special reference to effluents of sulfate pulp factories. Nord. Hyg. Tidskr. (Suppl. 2):1–182, 1970.

194. Lindvall, T. Sensory measurement of ambient traffic odors. J. Air Pollut. Control Assoc. 23:697–700, 1973.

195. Lindvall, T. Monitoring odorous air pollution in the field with human observers. Ann. N.Y. Acad. Sci. 237:247–260, 1974.

196. Lindvall, T., O. Norén, and L. Thyselius. On the abatement of animal manure odours, pp. E120–E123. In Proceedings of the 3rd International Clean Air Congress. Düsseldorf: VDI-Verlag, 1973.

197. Lindvall, T., and E. P. Radford, Eds. Measurement of annoyance due to exposure to environmental factors. Environ. Res. 6:1–36, 1973.

198. Lindvall, T., and L. T. Svensson. Equal unpleasantness matching of malodorous substances in the community. J. Appl. Psychol. 59:264–269, 1974.

199. Ljunggren, G., and B. Norberg. On the effect and toxicity of dimethyl sulfide, dimethyl disulfide and methyl mercaptan. Acta Phsyiol. Scand. 5:248–255, 1943.

200. Ljunggren, P. A sulfur mud deposit formed through bacterial transformation of fumarolic hydrogen sulfide. Econ. Geol. 55:531–538, 1960.

201. Lodge, J. P., Jr., J. B. Pate, B. E. Ammons, and G. A. Swanson. The use of hypodermic needles as critical orifices in air sampling. J. Air Pollut. Control Assoc. 16:197–200, 1966.

202. Lovelock, J. E., R. J. Maggs, and R. A. Rasmussen. Atmospheric dimethyl sulphide and the natural sulphur cycle. Nature 237:452–453, 1972.

203. Lund. O.-E., and H. Wieland. Pathologisch-anatomische Befunde bei experimenteller Schwefelwasserstoff vergiftung (H_2S). Eine Untersuchung an Rhesauaffen. Internes Arch. Gewerbe-pathol. Gewerbehyg. 22:46–54, 1966.

204. Lundgren, J. R., J. R. Vestal, and F. R. Tabita. The microbiology of mine drainage pollution, pp. 69–88. In R. Mitchell, Ed. Water Pollution Microbiology. New York: Wiley-Interscience, 1972.

205. Macaluso, P. Hydrogen sulfide, pp. 375–389. In H. F. Mark, J. J. McKetta, Jr., and D. F. Othmer. Kirk-Othmer Encyclopedia of Chemical Technology. (2nd ed.) New York: John Wiley & Sons, Inc., 1969.

206. Maroulis, P. J., and A. R. Bandy. Estimate of the contribution of biologically produced dimethyl sulfide to the global sulfur cycle. Science 196:647–648, 1977.

207. Martin, W., and A. C. Stern. The World's Air Quality Management Standards. Volume I. The Air Quality Management Standards of the World, Including United States Federal Standards. EPA-650/9-75-001-a. (Prepared for U.S. Environmental Protection Agency.) Chapel Hill: University of North Carolina, 1974. 382 pp.

208. Martin, W., and A. C. Stern. The World's Air Quality Management Standards. Volume II. The Air Quality Management Standards of the United States. EPA-650/9-75-001-b. (Prepared for U.S. Environmental Protection Agency.) Chapel Hill: University of North Carolina, 1974. 373 pp.

209. Masure, R. La kérato-conjonctivite des filatures de viscose; étude clinique et expérimentale. Rev. Belge de Path. 20:297–341, 1950.
210. McCabe, L. C., and G. D. Clayton. Air pollution by hydrogen sulfide in Poza Rica, Mexico. An evaluation of the incident of Nov. 24, 1950. A.M.A. Arch. Ind. Hyg. Occup. Med. 6:199–213, 1952.
211. McCallan, S. E. A., A. Hartzell, and F. Wilcoxon. Hydrogen sulphide injury to plants. Contrib. Boyce Thompson Inst. 8:189–197, 1936.
212. McCallan, S. E. A., and C. Setterstrom. Toxicity of ammonia, chlorine, hydrogen cyanide, hydrogen sulphide, and sulphur dioxide gases. I. General methods and correlations. Contrib. Boyce Thompson Inst. 11:325–330, 1940.
213. McCallan, S. E. A., and F. R. Weedon. Toxicity of ammonia, chlorine, hydrogen cyanide, hydrogen sulphide, and sulphur dioxide gases. II. Fungi and bacteria. Contrib. Boyce Thompson Inst. 11:331–342, 1940.
214. McDonald, R. Ophthalmological aspects of CS_2 intoxication, pp. 38–40. In Pennsylvania Department of Labor and Industry. Survey of Carbon Disulphide and Hydrogen Sulphide Hazards in the Viscose Rayon Industry. Bulletin No. 46. Occupational Disease Prevention Division. Harrisburg: Pennsylvania Department of Labor and Industry, 1938.
215. McKee, J. E., and H. W. Wolf, Eds. Water Quality Criteria. (2nd ed.) California: California State Water Resources Control Board, 1963.
216. Medvedev, V. M. The effect of certain industrial poisons on the mechanism of the nerve impulse transmission in the superior cervical sympathetic ganglion. Report I. Acute experiments with hydrogen sulphide, ethylene and propylene in healthy animals. Biul. Eksp. Biol. 47(4):79–82, 1959. (in Russian, summary in English)
217. Michel, H. O. A study of sulfhemoglobin. J. Biol. Chem. 126:323–348, 1938.
218. Milby, T. H. Hydrogen sulfide intoxication. Review of the literature and report of unusual accident resulting in two cases of nonfatal poisoning. J. Occup. Med. 4:431–437, 1962.
219. Miner, S. Preliminary Air Pollution Survey of Hydrogen Sulfide. A Literature Review. National Air Pollution Control Administration Publ. No. APTD 69-37. (Prepared for U.S. Department of Health, Education and Welfare.) Bethesda, Md.: Litton Systems, Incorporated, 1969. 91 pp. (Available from National Technical Information Service as Publ. No. PB-188 068.)
220. Mitchell, C. W., and S. J. Davenport. Hydrogen sulphide literature. Public Health Rep. 39:1–13, 1924.
221. Mitchell, C. W., and W. P. Yant. Correlation of the data obtained from refinery accidents with a laboratory study of H_2S and its treatment, pp. 59–80. In U.S. Bureau of Mines Bulletin 231. Washington, DC: U.S. Government Printing Office, 1925.
222. Modica, V., M. Rossi, and C. Sfogliano. Danni olfattivi da inalazione di vapori e fumi metallici. Clin. Otorinolaringoiatr. 16:416–425, 1964. (summary in English)
223. Moncrieff, R. W. Odour Preferences. London: Leonard Hill, 1966. 357 pp.
224. Moncrieff, R. W. Odours. London: William Heinemann Medical Books, Ltd., 1970. 237 pp.
225. Morell, D. B., Y. Chang, and P. S. Clezy. The structure of the chromophore of sulphmyoglobin. Biochim. Biophys. Acta 136:121–130, 1967.

226. Moskowitz, H. R., A. Dravnieks, W. S. Cain, and A. Turk. Standardized procedure for expressing odor intensity. Chem. Senses Flavor 1:235–237, 1974.

227. Moskowitz, H. R., A. Dravnieks, and L. A. Klarman. Odor intensity and pleasantness for a diverse set of odorants. Percept. Psychophys. 19:122–128, 1976.

228. Mower, G. D., R. G. Mair, and T. Engen. Influence of internal factors on the perceived intensity and pleasantness of gustatory and olfactory stimuli, pp. 103–121. In M. R. Kare and O. Maller, Eds. Chemical Senses and Nutrition. New York: Academic Press, 1977.

229. Nagel, R. L., and H. M. Ranney. Drug-induced oxidative denaturation of hemoglobin. Semin. Hematol. 10:269–278, 1973.

230. Nakamura, H. Über die Kohlensäureassimilation bei niederen Algen in Anwesenheit des Schwefelwasserstoffs. Acta Phytochim. 10:271–281, 1938.

231. National Academy of Sciences-National Academy of Engineering, Environmental Studies Board. Toxic substances, pp. 191–193. In Water Quality Criteria 1972. A report of the Committee on Water Quality Criteria. Washington, D.C.: U.S. Government Printing Office, 1972.

232. Ngo, T. T., and P. D. Shargool. The use of a sulfide ion selective electrode to study O-acetylserine sulfhydrylase from germinating rapeseed. Anal. Biochem. 54:247–261, 1973.

233. Ngo, T. T., and P. D. Shargool. The enzymatic synthesis of L-cysteine in higher plant tissue. Can. J. Biochem. 52:435–440, 1974.

234. Nichol, A. W., I. Hendry, D. B. Morell, and P. S. Clezy. Mechanism of formation of sulphaemoglobin. Biochim. Biophys. Acta 156:97–108, 1968.

235. Nicholls, P. The effect of sulphide on cytochrome aa_3. Isoteric and allosteric shifts of the reduced α peak. Biochim. Biophys. Acta 396:24–35, 1975.

236. Nyman, H. Th. Hydrogen sulfide eye inflammation—treatment with cortisone. Ind. Med. Surg. 23:161–162, 1954.

237. Oehme, F., and H. Wyden. Ein neues Gerät zur potentiometrischen Bestimmung kleiner Schwefelwasserstoffmengen in Luft und technischen Gasen. Staub 26(6):252, 1966.

238. O'Keeffe, A. E., and G. C. Ortman. Primary standards for trace gas analysis. Anal. Chem. 38:760–763, 1966.

239. O'Keeffe, A. E., and G. C. Ortman. Precision picogram dispenser for volatile substances. Anal. Chem. 39:1047, 1967.

240. Oliver, T. The sulphur miners of Sicily: Their work, diseases, and accident insurance. Brit. Med. J. 2:12–14, 1911.

241. Oseid, D. M., and L. L. Smith, Jr. Chronic toxicity of hydrogen sulfide to *Gammarus pseudolimnaeus*. Trans. Amer. Fish. Soc. 103:819–822, 1974.

242. Owen, H., and R. Gesell. Peripheral and central chemical control of pulmonary ventilation. Proc. Soc. Exp. Biol. Med. 28:765–766, 1931.

243. Paez, D. M., and O. A. Guagnini. Isolation and ultramicro determination of hydrogen sulfide in air and water by use of ion-exchange resin. Mikrochim. Acta (2):220–224, 1971.

244. Pantaleoni, R. Critique of communication entitled the enzyme model of olfaction, dual nature of odorivectors and specific malodor counteractants by Alfred A. Schleppnik, Monsanto Flavor/Essence. Perfume. Flavor. 1(5):16–17, 1976.

245. Paré, J. P. A new tape reagent for the determination of hydrogen sulfide in air. J. Air Pollut. Control Assoc. 16:325–327, 1966.

246. Parker, C. D. The corrosion of concrete. 1. The isolation of a species of bacterium associated with the corrosion of concrete exposed to atmospheres containing hydrogen sulphide, pp. 81–90. 2. The function of *Thiobacillus concretivorus* (Nov. spec.) in the corrosion of concrete exposed to atmospheres containing hydrogen sulphide. pp. 91–98. Aust. J. Exp. Biol. Med. Sci. 23:81–98, 1945.
247. Partington, J. R. Chemistry in Scandinavia, II. Scheele, pp. 205–234. In A History of Chemistry. Vol. 3. London: Macmillan & Co., Ltd., 1962.
248. Pecsar, R. E., and C. H. Hartmann. Automated gas chromatographic analysis of sulfur pollutants. Anal. Instrum. 9:H-2-1–H-2-14, 1971.
249. Peisach, J. An interim report on electronic control of oxygenation of heme proteins. Ann. N.Y. Acad. Sci. 244:187–203, 1975.
250. Petrun, N. M. Signs indicative of hydrogen sulfide poisoning following its entry into the organism via the skin. Farmakol. Toxshikol. 28:488–490, 1965. (in Russian, summary in English)
251. Pfaffmann, C. The pleasures of sensation. Psychol. Rev. 67:253–268, 1960.
252. Pitts, G., A. I. Allam, and J. P. Hollis. *Beggiatoa*: Occurrence in the rice rhizosphere. Science 178:990–991, 1972.
253. Poda, G. A. Hydrogen sulfide can be handled safely. Arch. Environ. Health 12:795–800, 1966.
254. Ramazini, B. De Morbis Artificum Diatriba. Typis: Antonii Capponi, 1700. 360 pp. (in Latin)
255. Ramazzini, B. Diseases of Workers. (Translated from the Latin text De Morbis Artificum of 1713 by W. C. Wright.) New York: Hafner Publishing Company, 1964. 549 pp.
256. Rankine, B. C. Hydrogen sulphide production by yeasts. J. Sci. Food Agric. 15:872–877, 1964.
257. Rankine, D. Artificial silk keratitis. Brit. Med. J. 2:6–9, 1936.
258. Reynolds, T. Effects of sulphur-containing compounds on lettuce fruit germination. J. Exp. Bot. 25:375–389, 1974.
259. Robinson, E., and R. C. Robbins. Gaseous sulfur pollutants from urban and natural sources. J. Air Pollut. Control Assoc. 20:233–235, 1970.
260. Rubin, H., and A. Arieff. Carbon disulfide and hydrogen sulfide: Clinical study of chronic low-grade exposures. J. Ind. Hyg. Toxicol. 27:123–129, 1945.
261. Ruch, W. E. Quantitative Analysis of Gaseous Pollutants, p. 131. Ann Arbor: Humphrey Science Publishers, Inc., 1970.
262. Saito, E., K. Wakasa, M. Okuma, and G. Tamura. Studies on the sulfite reducing system of algae. Part III. Sulfite reduction by algal extract coupling to the reduced ferredoxin. Bull. Assoc. Nat. Sci. Senshu Univ. 3:45–50, 1970.
263. Saltzman, B. E. Direct reading colorimetric indicators, pp. S-3–S-10. In Air Sampling Instruments for Evaluation of Atmospheric Contaminants (4th ed.) Cincinnati: American Conference of Governmental Industrial Hygienists, 1972.
264. Saltzman, H. A., and H. O. Sieker. Intestinal response to changing gaseous environments: Normobaric and hyperbaric observations. Ann. N.Y. Acad. Sci. 150:31–39, 1968.
265. Sandage, C., and K. C. Back. Effects of animals of 90-day continuous inhalation exposure to toxic compounds. Fed. Proc. 21:451, 1962. (abstr.)
266. Sanderson, H. P., R. Thomas, and M. Katz. Limitations of the lead acetate impregnated paper tape method for hydrogen sulfide. J. Air Pollut. Control Assoc. 16:328–330, 1966.

267. Sandusky, A., and A. Parducci. Pleasantness of odors as a function of the immediate stimulus context. Psychonomic Sci. 3:321–322, 1965.

268. Sayers, R. R., C. W. Mitchell, and W. P. Yant. Hydrogen Sulphide as an Industrial Poison. U.S. Bureau of Mines, Department of the Interior. Reports of Investigations, Serial No. 2491. Washington, D.C.: U.S. Department of the Interior, 1923. 6 pp.

269. Sayers, R. R., N. A. C. Smith, A. C. Fieldner, C. W. Mitchell, G. W. Jones, W. P. Yant, D. D. Stark, S. H. Katz, J. J. Bloomfield, and W. A. Jacobs. Investigation of Toxic Gases from Mexican and Other High-Sulphur Petroleums and Products. Bulletin No. 231. Report by the Department of the Interior, Bureau of Mines, to the American Petroleum Institute. Washington, D.C.: U.S. Government Printing Office, 1925. 108 pp.

270. Scaringelli, F. P., S. A. Frey, and B. E. Saltzman. Evaluation of Teflon permeation tubes for use with sulfur dioxide. Amer. Ind. Hyg. Assoc. J. 28:260–266, 1967.

271. Scaringelli, F. P., A. E. O'Keeffe, E. Rosenberg, and J. P. Bell. Preparation of known concentrations of gases and vapors with permeation devices calibrated gravimetrically. Anal. Chem. 42:871–876, 1970.

272. Scheler, W. Zur Komplexaffinität von Hämoglobine. Z. Physik. Chem. 210:61–71, 1959.

273. Scheler, W., and R. Kabisch. Über die antagonistische Beeinflussung der akuten H_2S-Vergiftung bei der Maus durch Methämoglobinbildner. Acta Biol. Med. Ger. 11:194–199, 1963.

274. Schiffman, S. S. Physiochemical correlates of olfactory quality. Science 185:112–117, 1974.

275. Schnyder, J., and K. H. Erismann. Zum Vorkommen von Thiorthreonin in Pflanzenmaterial. Experientia 29:232, 1973.

276. Schnyder, J., M. Rottenberg, and K. H. Erismann. The synthesis of threonine and thiothreonine from *o*-phospho-homoserine by extracts prepared from higher plants. Biochem. Physiol. Pflanz. 167:605–608, 1975.

277. Seltzer, R. J. Monsanto develops malodor counteractant. Chem. Eng. News 53(41):24–25, 1975.

278. Shigeta, Y. Research on odor abatement and control in U.S.A. (II). Akushu no Kenkyu (Odor Res. J. Jap.) 1(4):9–20, 1971. (in Japanese)

279. Silverman, M. P., and H. L. Ehrlich. Microbial formation and degradation of minerals. Adv. Appl. Microbiol. 6:153–206, 1964.

280. Simson, R. E., and G. R. Simpson. Fatal hydrogen sulphide poisoning associated with industrial waste exposure. Med. J. Aust. 1:331–334, 1971.

281. Singer, A. G., W. C. Agosta, R. J. O'Connell, C. Pfaffmann, D. V. Bowen, and F. H. Field. Dimethyl disulfide: An attractant pheromone in hamster vaginal secretion. Science 191:948–950, 1976.

282. Slater, E. C. The components of the dihydrocozymase oxidase system. Biochem. J. 46:484–499, 1950.

283. Smith, L., H. Kruszyna, and R. P. Smith. The effect of methemoglobin on the inhibition of cytochrome *c* oxidase by cyanide, sulfide or azide. Biochem. Pharmacol. 26:2247–2250, 1977.

284. Smith, R. P. The oxygen and sulfide binding characteristics of hemoglobins generated from methemoglobin by two erythrocytic systems. Mol. Pharmacol. 3:378–385, 1967.

285. Smith, R. P. Cobalt salts: Effects in cyanide and sulphide poisoning and on methemoglobinemia. Toxicol. Appl. Pharmacol. 15:505–516, 1969.

286. Smith, R. P., and R. A. Abbanat. Protective effect of oxidized glutathione in acute sulfide poisoning. Toxicol. Appl. Pharmacol. 9:209–217, 1966.

287. Smith, R. P., and R. E. Gosselin. The influence of methemoglobinemia on the lethality of some toxic anions. II. Sulfide. Toxicol. Appl. Pharmacol. 6:584–592, 1964.

288. Smith, R. P., and R. E. Gosselin. On the mechanism of sulfide inactivation by methemoglobin. Toxicol. Appl. Pharmacol. 8:159–172, 1966.

289. Smith, R. P., R. Kruszyna, and H. Kruszyna. Management of acute sulfide poisoning: Effects of oxygen, thiosulfate, and nitrite. Arch. Environ. Health 31:166–169, 1976.

290. Smith, R. P., and M. V. Olson. Drug-induced methemoglobinemia. Semin. Hematol. 10:253–268, 1973.

291. Smith, R. P., and C. D. Thron. Hemoglobin, methylene blue and oxygen interactions in human red cells. J. Pharmacol. Exp. Ther. 183:549–558, 1972.

292. Sörbo, B. On the formation of thiosulfate from inorganic sulfide by liver tissues and heme compounds. Biochim. Biophys. Acta 27:324–329, 1958.

293. Sörbo, B. On the mechanism of sulfide oxidation in biological systems. Biochim. Biophys. Acta 38:349–351, 1960.

294. Springer, K. J., and C. T. Hare. A Field Survey to Determine Public Opinion of Diesel Engine Exhaust Odor. Final Report on Contract PH-22-68-36. SwRI-AR-718. San Antonio, Tex.: Southwest Research Institute, 1970. 166 pp.

295. Stella, G. The reflux response of the "apneustric" centre to stimulation of the chemoreceptors of the carotid sinus. J. Physiol. 95:365–372, 1939.

296. Stern, A. C., Ed. Air pollution standards, p. 602. Air Pollution. Vol. III. (2nd ed.) New York: Academic press, 1968.

297. Stern, K. G. Über die Hemmungstypen und den Mechanismus der katalatischen Reaktion. 3. Mitteilung über Katalase. Hoppe-Seyler's Z. Physiol. Chem. 209:176–206, 1932.

298. Stevens, R. K., J. D. Mulik, A. E. O'Keeffe, and K. J. Krost. Gas chromatography of reactive sulfur gases in air at the parts-per-billion level. Anal. Chem. 43:827–831, 1971.

299. Stevens, R. K., and A. E. O'Keeffe. Modern aspects of air pollution monitoring. Anal. Chem. 42:143A-148A, 1970.

300. Stevens, S. S. On the psychophysical law. Psychol. Rev. 64:153–181, 1957.

301. Stine, R. J., B. Slosberg, and B. E. Beacham. Hydrogen sulfide intoxication: A case report and discussion of treatment. Ann. Intern. Med. 85:756–758, 1976.

302. Subramoney, N. Injury to paddy seedlings by production of H_2S under field conditions J. Indian Soc. Soil Sci. 13:95–98, 1965.

303. Svensson, L. T., and T. Lindvall. On the consistency of intramodal intensity matching in olfaction. Percept. Psychophys. 16:264–270, 1974.

304. Taras, M. J., A. E. Greenberg, R. D. Hoak, and M. C. Rand, Eds. Methylene blue visual color-matching method, pp. 555–558. In Standard Methods for the Examination of Water and Wastewater. (13th ed.) Washington, D.C.: American Public Health Association, 1971.

305. Tentative method of analysis for hydrogen sulfide content of the atmosphere, pp. 426–432. In Methods of Air Sampling and Analysis. Washington, D.C.: American Public Health Association, 1972.

306. Theede, H., A. Ponat, K. Hiroki, and C. Schlieper. Studies on the resistance of marine bottom invertebrates to oxygen-deficiency and hydrogen sulphide. Mar. Biol. 2:325–337, 1969.

307. Third Karolinska Institute Symposium on Environmental Health. Methods for measuring and evaluating air pollutants at the source and in the ambient air. Report of an international symposium in Stockholm, June 1–5, 1970. Nord. Hyg. Tidskr. 51(2):1–77, 1970.

308. Thoen, G. N., G. G. DeHaas, and R. R. Austin. Instrumentation for quantitative measurement of sulfur compounds in kraft gases. Tappi 51:246–249, 1968.

309. Thompson, J. F., and D. P. Moore. Enzymatic synthesis of cysteine and S-methylcysteine in plant extracts. Biochem. Biophys. Res. Commun. 31:281–286, 1968.

310. Thornsberry, W. L., Jr. Isothermal gas chromatographic separation of carbon dioxide, carbon oxysulfide, hydrogen sulfide, carbon disulfide, and sulfur dioxide. Anal. Chem. 43:452–453, 1971.

311. Thornton, N. C., and C. Setterstrom. Toxicity of ammonia, chlorine, hydrogen cyanide, hydrogen sulphide, and sulphur dioxide gases. III. Green plants. Contrib. Boyce Thompson Inst. 11:343–356, 1940.

312. Tucker, D. Nonolfactory responses from the nasal cavity: Jacobson's organ and the trigeminal system, pp. 151–181. In L. M. Beidler, Ed. Handbook of Sensory Physiology. Vol. 4. Berlin: Springer-Verlag, 1971.

313. Tuovinen, O. H., and D. P. Kelly. Biology of *Thiobacillus ferrooxidans* in relation to the microbiological leaching of sulphide ores. Z. Allg. Mikrobiol. 12:311–346, 1972.

314. Turrell, F. M. The area of the internal exposed surface of dicotyledon leaves. Amer. J. Bot. 23:255–264, 1936.

315. Union of International Associations. World Health Organization, p. 645. In Yearbook of International Organizations. (15th ed.) Brussels, Belgium: Union of International Associations, 1974.

316. U.S. Department of Health, Education and Welfare. Air Quality Criteria for Particulate Matter. Air Pollution Control Administration Publ. No. AP-49. Washington, D.C.: U.S. Government Printing Office, 1969. 211 pp.

317. U.S. Department of Health, Education and Welfare. Air Quality Criteria for Carbon Monoxide. Air Pollution Control Administration Publ. No. AP-62, 1970. [184 pp.]

318. U.S. Department of Health, Education and Welfare. Air Quality Criteria for Hydrocarbons. Air Pollution Control Administration Publ. No. AP-64. Washington, D.C.: U.S. Government Printing Office, 1970. 126 pp.

319. U.S. Department of Health, Education and Welfare. Air Quality Criteria for Photochemical Oxidants. Air Pollution Control Administration Publ. No. AP-63. Washington, D.C.: U.S. Government Printing Office, 1970. 202 pp.

320. U.S. Department of Health, Education and Welfare. Air Quality Criteria for Sulfur Oxides. Air Pollution Control Administration Publ. No. AP-50. Washington, D.C.: U.S. Government Printing Office, 1970. 177 pp.

321. U.S. Department of Health, Education, and Welfare. National Institute for Occupational Safety and Health. Division of Laboratories and Criteria Development. NIOSH Manual of Analytical Methods. NIOSH 75-121, p. 126. Cincinnati: U.S. Department of Health, Education and Welfare, 1974.

322. U.S. Department of Labor, Occupational Safety and Health Administration. Table G-3, Mineral dusts, p. 23543. In Chapter XVII, Occupational safety and health standards. Fed. Reg. 39:June 27, 1974.

323. U.S. Environmental Protection Agency. Air Quality Criteria for Nitrogen

Oxides. Air Pollution Control Office Publ. No. AP-84. Washington D.C.: U.S. Government Printing Office, 1971. [188 pp.]

324. U.S. Public Health Service, National Institute of Health, Division of Industrial Hygiene. Hydrogen sulfide: Its toxicity and potential dangers. Pub. Health Rep. 56:684–692, 1941.

325. Verein Deutscher Ingenieure. Schwefelwasserstoff. Maximale Immissions-Konzentrationen. (MIK-Werte) VDI 2107, 7 pp. In VDI-Handbuch Reinhaltung der Luft. Band I. Düsseldorf: VDI-Verlag GmbH, 1960.

326. Verein Deutscher Ingenieure. Gasauswurfbegrenzung. Schwefeldioxyd. Kokereien und Gaswerke. Koksöfen (Abgase). VDI 2110, 8 pp. In VDI-Handbuch Reinhaltung der Luft. Band I. Düsseldorf: VDI-Verlag GmbH, 1960.

327. Von Wolzogen Kühr, C. A. H., and L. S. Van der Vlugt. Aerobic and anaerobic iron corrosion in water mains. J. Amer. Water Works Assoc. 45:33–46, 1953.

328. Waller, R. L. Methanethiol inhibition of mitochondrial respiration. Toxicol. Appl. Pharmacol. 42:111–117, 1977.

329. Wallis, R. L. M. On sulphaemoglobinaemia. Q. J. Med. 7:176–206, 1913–1914.

330. Walton, D. C., and M. G. Witherspoon. Skin absorption of certain gases. J. Pharmacol. Exp. Ther. 26:315–324, 1925.

331. Weast, R. C., Ed. CRC Handbook of Chemistry and Physics. (57th ed.) Cleveland: Chemical Rubber Company Press, 1976. 2390 pp.

332. Weedon, F. R., A. Hartzell, and C. Stetterstrom. Toxicity of ammonia, chlorine, hydrogen cyanide, hydrogen sulphide, and sulphur dioxide gases. V. Animals. Contrib. Boyce Thompson Inst. 11:365–385, 1940.

333. Wieland, H., and H. Sutter. About oxidases and peroxidases. Chem. Abstr. 22:2574–2575, 1928.

334. Winder, C. V., and H. O. Winder. The seat of action of sulfide on pulmonary ventilation. Amer. J. Physiol. 105:337–352, 1933.

335. Windholz, M., Ed. Hydrogen sulfide, pp. 633–634. In The Merck Index. (9th ed.) Rahway, N.J.: Merck & Co., Inc., 1976.

336. Winek, C. L., W. D. Collom, and C. H. Wecht. Death from hydrogen-sulphide fumes. Lancet 1:1096, 1968.

337. Woskow, M. H. Multidimensional scaling of odors, pp. 147–188. In N. N. Tanyolac, Ed. Theories of Odor and Odor Measurement. Istanbul: Robert College, 1968.

338. Yant, W. P. Hydrogen sulphide in industry. Occurrence, effects and treatment. Amer. J. Public Health 20:598–608, 1930.

339. Yoshida, M. Studies in psychometric classification of odors. (4). Jap. Psychol. Res. 6:115–124, 1964.

340. Zajic, J. E. Microbial Biogeochemistry. New York: Academic Press, 1969. 345 pp.

341. Zieve, L., W. M. Doizaki, and F. J. Zieve. Synergism between mercaptans and ammonia or fatty acids in the production of coma: A possible role for mercaptans in the pathogenesis of hepatic coma. J. Lab. Clin. Med. 83:16–28, 1974.

Index